T0191692

Behavioral Issues in Operations Management

Ilaria Giannoccaro
Editor

Behavioral Issues in Operations Management

New Trends in Design, Management, and Methodologies

 Springer

Editor
Ilaria Giannoccaro
Department of Mechanics, Mathematics, and Management
Polytechnic University of Bari
Bari
Italy

ISBN 978-1-4471-5807-3   ISBN 978-1-4471-4878-4 (eBook)
DOI 10.1007/978-1-4471-4878-4
Springer London Heidelberg New York Dordrecht

Foreword

In recent years, we have witnessed an emerging area of research termed behavioral operations management. Major academic journals have published special issues, and professional societies have held conferences to communicate the exciting developments within this new field. While there have been previous books written on the topic, the current book marks an important step forward in establishing this genre of academic research.

This book focuses on the spectrum of contexts ranging from individual and organizational behaviors to the behavior of networks of organizations. It addresses decision making, cognitive biases, cultural norms, organization knowledge, and politics at the individual, organizational, and network levels. The topics of decentralized decision making and local autonomy are considered, along with the issues of power, conflict, trust, and equity. The recurring themes throughout this book are the ways to positively affect the process and system dynamics across different units of analysis. Though it is not intended to be a casual read, this is a great reference book offering many ideas for future research and methodologies pertaining to the exciting field of behavioral operations management.

Thomas Y. Choi
Bob Herberger Arizona Heritage Chair
Director, Center for Supply Networks
Arizona State University

Contents

Introduction

Behavioural Operations Management (BOM) has been identified in the last years as one of the most promising emerging fields in Operations Management (OM) (Bendoly et al. 2006; Gino and Pisano 2008; Loch and Wu 2007). BOM explicitly studies the effects of human behaviour on the performances of operating systems and analyses strategies to improve them (Gino and Pisano 2008; Loch and Wu 2007). In particular, BOM explores deviations from rationality of the decision makers involved in the management of operating systems including factors affecting their behavior (Siemsen 2009), with the aims firstly of providing a better understanding of how operating systems work and perform, and secondly of developing effective implications for the design, management, and improvement of operating systems (Gino and Pisano 2008).

Two main aspects characterize BOM research: one is referring to the choice of the appropriate research methodology and concerns the unresolved tension between modelling and empirical studies; the other regards its multi-disciplinary nature, encompassing many different disciplines among which organizational behavior, decision science, behavioral decision making, psychology, and management.

The book is organized to account for these two aspects with the two-fold aim to frame the state of the art of the field and to offer innovative contributions and inspirations for moving beyond the traditional issues. Therefore, the book has been thought to provide an update on the established research methodologies as well as some suggestions for the application of new methodologies particularly promising for the topic. Furthermore, I have collected studies in various fields authored by leading scholars coming from different areas so as to offer an extended view of the behavioral factors influencing OM.

The book comprises 11 chapters. The first two chapters explicitly address the methodological aspect. In the chapter *"Behavioural OM Experiments: Critical Inquiry Reawakening Practical issues in Research"*, Bendoly and Eckerd outline the opportunities made available in the OM literature through experimental behaviors. They analyze the different options available to experimental design, i.e., vignette, process simulation, and experimental design, and discuss the main features, the benefits, and the limitations of each method. They conclude dealing with the most interesting contributions resulting from experimentation and suggesting future research directions.

In the chapter *"Complex Systems Methodologies for Behavioural Research in Operations Management: NK Fitness Landscape"*, Giannoccaro expands the traditional set of research methodologies used in BOM by including a complexity science tool, i.e., NK fitness landscape. She first discusses the opportunities for doing successful BOM research made available by adopting complexity science and then explains how NK fitness landscape may be employed to simulate different OM contexts and which research questions may be addressed.

The subsequent contributors examine how specific behavioral factors (trust, cognitive capacity, and motivation) affect the design and management of complex operating systems at the various dimensional levels, i.e., shopfloor, firm, and network of firm. How these factors affect decision making in multiple OM contexts are also investigated, including logistics, supply chain management, purchasing, and human resource management.

In particular, two papers analyze the role of trust. In the chapter *"Trust in Face-To-Face and Electronic Negotiation in Buyer–Supplier Relationships: A Laboratory Study"*, Moramarco, Stevens, and Pontrandolfo investigate whether having trusting relationships with the suppliers can positively affect outcomes and strengthen the relationship even when electronic mechanisms are used for purchasing. They conduct a laboratory study which compares three negotiation mechanisms (i.e., face-to-face negotiation, e-mail negotiation, and e-reverse auction).

In the chapter *"Lean Supply Chain: A behavioural Perspective: Examples from Packaging Supply Chains in the FMCG Sector"*, Found highlights the importance of trust, power, and equity in implementing JIT operations and lean relationship management. By conducting a case study in the packaging sector, she shows that trust is an important element to sustain high-performance JIT operations and lean relationships management and that strong supply relationships based on mutual trust and equity are a prerequisite for a successful implementation of lean supply chains.

A further behavioral factor addressed in the book is related to the cognitive limitations of the decision makers. In the chapter *"Supply Chain Integration: A Behavioural Study Using NK Simulation"*, Giannoccaro investigates the extent to which the cognitive abilities and the resistance to change of the decision maker influences the effectiveness of an integrated management approach of the supply chain. She conducts a simulation study using the NK fitness landscape and shows that the complexity of the supply chain is a factor strengthening the impact of the decision maker behavior on the performance.

In the chapter "*Cognitive Biases, Heuristics and Overdesign: An Investigation on the Unconscious Mistakes of Industrial Designers and on their Effects on Product Offering*" by Belvedere, Grando, and Ronen, the effects of diverse cognitive biases of decision makers are investigated in the context of product design. The casual relationship between cognitive biases and overdesign is then postulated and tested.

The problem of motivation and of the accurate design of the incentives for stimulating more effective decision maker behavior is investigated by two other papers.

In the chapter "*Incentives in Organizations of Operating Systems: Can Economics and Psychology Coexist in Human Resources Management?*", Merlone highlights that human behavior should be included for designing successful contracts. By illustrating how two disciplines, Economics and Psychology, address Human Resource Management, he suggests interesting approaches for building effective contracts also dealing with complexity.

In the chapter "*Incentives for Cost Transparency Implementation: A Framework from an Action Research*", Romano e Formentini conduct an action research aimed at developing a framework useful to identify the appropriate forms of incentives to stimulate suppliers to share cost information.

The last three contributions extend the traditional OM contexts in which the effects of the behavioral issues are usually analyzed. In the chapter "*Learning on the Shop Floor: The Behavioural Roots of Organisational Knowledge*", Hanson proposes an interesting behavioral view of the organizational knowledge, which is particularly useful for its implications on knowledge management and learning.

In the chapter "*Behavioural Decision-Making and Network Dynamics: A Political Perspective*", Zirpoli, Errichiello and Whitford shed light on the mechanisms underlying the functioning of the network as a form of governance by proposing a theoretical model in which the network is view as a political coalition. They underline the central role played by power and politics in shaping the firm' organizational boundaries. Their model is particularly interesting for its implications on vertical integration and outsourcing.

Finally, in the chapter "*Markets of Logistics Services: The Role of Actors' Behaviour to Enhance Performance*", Bellantuono, Kersten, and Pontrandolfo develop a simulation model of the decision-making process in a logistic system organized as a market with the aim to identify appropriate strategies to enhance coordination among actors. They show that such strategies recommending to *select, enrich,* and *modify* information exchanged by actors need to be designed and implemented by taking into account behavioral issues related to the subjective social perceptions and the expectations of the decision makers regarding themselves, the counterpart, and the context in which the transaction occurs.

Ilaria Giannoccaro

References

E. Bendoly, K. Donohue, K. Schultz, Behavior in operations management: assessing recent findings and revisiting old assumptions. J. Oper. Manag. **24**, 737–752 (2006)

F. Gino, G. Pisano, Toward a theory of behavioural operations. Manuf. Serv. Oper. Manag. **10**, 676–691 (2008)

C.H. Loch, Y. Wu, *Behavioral Operations Management* (Now Publishers Inc., Hanover, 2007)

E. Siemens, That thing called Be-OP's. POMS Chronicle **16**(1), 12–13 (2009)

Chapter 1
Behavioral OM Experiments: Critical Inquiry Reawakening Practical Issues in Research

Elliot Bendoly and Stefanie Eckerd

Abstract Behavioural Operations Management is a multi-disciplinary branch of OM that explicitly considers the effects of human behaviour on process and system dynamics, influenced by cognitive biases and limitations, social preferences and perceptions of cultural norms. Conversely this domain also concerns itself with the effect of process and system dynamics on human behaviour, hence viewing human behaviour as critical in not only its direct and moderating effects but also in its mediating role between operating policy change and connected outcomes. In this chapter we discuss insights from the various contributing research literatures, as well as an overview of theory and methods applied. We cite the most influential papers in the Behavioural Operations Management literature in our discourse.

1 An Overview

Let us start with a definition...

Behavioral Operations Management: A multidisciplinary branch of OM that explicitly considers the effects of human behavior on process and system dynamics, influenced by cognitive biases and limitations, social preferences and perceptions of cultural norms. Conversely this domain also concerns itself with the effect of process and system

E. Bendoly (✉)
Goizueta Business School, Emory University, 1300 Clifton Road NE,
Atlanta GA 30322, USA
e-mail: Elliot_Bendoly@bus.emory.edu

S. Eckerd
Robert H. Smith School of Business, Logistics, Business and Public Policy,
3341 Van Munching Hall, College Park MD 20742-1815, USA
e-mail: seckerd@rhsmith.umd.edu

I. Giannoccaro (ed.), *Behavioral Issues in Operations Management*,
DOI: 10.1007/978-1-4471-4878-4_1, © Springer-Verlag London 2013

dynamics on human behavior, hence viewing human behavior as critical in not only its direct and moderating effects but also in its mediating role between operating policy change and connected outcomes.

This definition is a variation on that proposed by earlier authors. In particular, this definition views the OM field's interest in human behavior not only from both an inputs as well as an outcomes perspective, but also from the perspective of behavioral processes of filtering and interpretation. Unlike the definition of Loch and Wu (2007), the above definition does not limit consideration to the testing of mathematical theory, which would be perhaps a more appropriate definition for Behavioral Operations Research.[1] Operations Management as a field is not a methods-focused discipline like OR, but rather a management domain focused field. Methods are merely a means to an end in OM, selected to match the management issue faced in practice. For the same reason, while certain methods have been particularly popular in Behavioral OM studies (c.f. Bendoly et al. 2006), the branch of Behavioral Operations Management is not restricted to any one method or theoretical foundation.

Specifically, while it may seem convenient to be critical of the long tradition of normative modeling (much of the work in OR) in terms of the extent of assumptions made regarding human actors in operations contexts, criticism can also be leveled against a range of empirical factor models in OM work that have not sufficiently considered the human element. To be sure, a large number of Behavioral OM studies have leveraged normative models in particular in an attempt to demonstrate the problematic nature of modeling assumptions and further provide insight into why so many of these models fall short in terms of their effective practical prescription. However, Behavioral OM also has a great potential to shore up various gaps between the prescriptions of existing empirically buttressed variance studies and practice. As Bendoly et al. (2010a) suggest in their discussion of the bodies of knowledge that feed into theoretical considerations in Behavioral OM studies, a wide array of disciplines can be drawn upon toward filling these gaps and fulfilling the mission of the above definition.

Having made this generalization regarding Behavioral OM's nonbinding relationship with specific research methods, it is nevertheless important to discuss the virtues of various research methods that can be leveraged in order to advance the field. This is after all one of the motivations behind the present text. In this chapter, we will specifically focus on the opportunities made available through one particular method of inquiry and its variants: experimental behavioral studies. We will outline not only the tradition of various management topics, experimental research methods applied, and foundational disciplinary theory drawn upon in the Behavioral OM literature to date, but also provide recommendations for future work aimed at capitalizing on the findings of existing work.

---

[1] For contrast between references to Behavioral OR and Behavioral OM, see, respectively, popularized Wiki definitions managed by Bearden (2010) and Eckerd and Bendoly (2010).

2 MultiDisciplinary Insights and Critical Calls

It has long been understood that humans are limited in their ability to collect and process information. When making decisions, especially complex decisions, human decision makers fail to adhere to normative decision theories, but interestingly, appear to do so in systematic ways (Kahneman et al. 1982). Moreover, a person's social goals and collective behaviors impart clear influences on behavior. Theories that are fundamental to the areas of cognitive psychology, social psychology, and sociology offer rich insights into the phenomena observed in operations management. They also provide guidance as to how behavioral experimental methods might be leveraged in OM contextual studies.

2.1 Cognitive Psychology

The cognitive revolution in psychology was important because it recognized an 'operant' individual acting between a stimulus and a response, capable of moderating the relationships between stimuli and responses which were previously believed to be mechanistic (Seligman and Maier 1967). Psychological and Organizational Behavior models (see below) had to be developed to account for unobservable, affective, and seemingly irrational responses from individuals. In particular, cognitive psychology addresses (among other things) an individual's decision-making biases and use of heuristics as an attempt to overcome bounded rationality. Heuristics are mapped to deviations in the decision-making process, and often lead to biases that are mapped to deviations in decision outcomes (Bendoly et al. 2010a, b). The anchoring and insufficient adjustment heuristic falls under this domain, and is employed when people attempt to estimate unknown data points. In an operations management context, orders for inventory may be anchored on mean demand and then insufficiently adjusted toward the normative order quantity (Schweitzer and Cachon 2000). Other behavioral regularities falling within the realm of cognitive psychology include framing effects and the overconfidence effect.

Since operations management often involves accounting for individual decision-making, or actions within contexts subject to OM design and policy, it is absolutely imperative to understand the potential impact that cognition and psychological phenomena have on these decisions and actions. Operations management researchers have only begun to realize through the use of behavioral experiments (c.f. Bendoly and Cotteleer 2008; Bendoly and Prietula 2008) how ignoring the existence of behavioral dynamics undermines the tenability of management research prescriptions for practice.

2.2 Social Psychology

Social psychology describes how an individual relates to other individuals, and specifically how individuals' actions are influenced by emotions (Loch and Wu 2007) and motivation (Bendoly et al. 2010a, b). Social behavioral theories help us understand why individuals act competitively or cooperatively with others. For example, those seeking status make decisions consistent with the achievement of recognition or higher hierarchical position relative to peers as an end goal. Status seeking as a social preference in operations management is observed in laboratory experiments, where subjects are shown to be willing to sacrifice supply chain profits and efficiency in response to aggressive pricing by their supply chain partner; in other words, they are willing to forfeit their own profits to prevent the aggressor from achieving status (Loch and Wu 2008). In addition to status, important social psychology facets include goal setting, feedback and controls, interdependence, and reciprocity.

Since operations management contexts seldom involve individuals acting in true isolation from others, it is entirely reasonable to assume their actions may in some way be influenced by their social as well as operational task settings. Here, behavioral experiments can be crucial in distinguishing task-specific and social context-specific socio-psychological phenomena associated with different operations management policies.

2.3 Sociology and Systems

Sociological theories define the context of interactions between individuals and groups, as well as the interactions between multiple groups, sometimes referred to as group dynamics. The concept of groupthink fits within this body of knowledge, wherein one individual changes her beliefs to conform to the larger group consensus. A strong group identity, and associated group-think, is a common point in team life cycles, and can prevent teams from accepting outside advice and incorporating external ideas. In operations management, product development teams can fall prey to this phenomenon and thus stall in creative and innovative efforts. Examinations of organizational and national cultural variations are important facets of work in this area as well.

Since much of modern OM practice hinges on communication, cooperation, and in some cases explicit collaboration, the role of individuals in interaction with each other may be highly relevant in the translation of operations management policy to performance. In order to truly understand group dynamics, they must be studied through experimentation where the simultaneous observation of multiple player actions can be recorded and analyzed for potential causal linkages, feedback structures, and overall system dynamics.

3 Types of Experimental Behavioral Studies

> The design of an experiment to test a particular theory often forces the experimenter to focus on specific aspects of the theory other than those that naturally come to the fore in the theoretical literature. The insights gained from designing an experiment are often of value even apart from the actual conduct of the experiment. Thus there is an interplay, on many levels, between theory and experiment—Kagel and Roth (1995, p. 10).

As stated by Kagel and Roth (1995) in their seminal work, the various linkages between theoretical argumentation and experimental design are inextricable. Any experimental design must be motivated and justified by the theory housing the research questions it intends to examine. Even the most rigorously conducted experiments may yield fruitless results, if the more fundamental choice of methodological approach does not connect clearly with the research's core theoretical model. Having said this, it is useful to consider some general options available to experimental design.

3.1 Vignettes

One classical approach to the study of human behavior is the use of static descriptive vignettes. Vignettes are useful for evaluating the intended reasoning, decision-making processes, and/or the intended behaviors of respondents. While they have been predominantly used in the realms of business ethics, marketing, public policy, and healthcare, to name a few, research employing the vignette methodology is finding a foothold in the OM literature in recent years (Mantel et al. 2006). As various disciplines have undertaken their own independent development of this methodology, studies employing this technique use different nomenclatures to describe it, including for example, scenarios, policy-capturing, stated-choice method, conjoint analysis, and contingent valuation method (Caro et al. 2010). Vignettes may be broadly defined as "short descriptions of a person or social situation which contain precise references to what are thought to be the most important factors in the decision-making or judgment-making processes of respondents" (Alexander and Becker 1978, p. 94). Three key components of a vignette experiment include, as suggested in the definition, a decision scenario that provides a standardized stimulus context, manipulated critical variables of interest, and response items.

The decision scenario depicts the context under study, and is presented as "focused descriptions of a series of events taken to be representative, typical or emblematic of a case/situation" (Ashill and Yavas 2006, p. 28). The scenario provides respondents with a tangible situation that helps make complex processes understandable, and a standardized stimulus so that respondents are restricted to a common field of vision and are distanced from their own personal experiences (Frederickson 1986; Finch 1987). Decision scenarios serve to maintain uniformity and control through consistent chronological flow of events and limitation of the time, actors, and space involved (Alexander and Becker 1978; Ashill and Yavas 2006). Those scenarios that are most successful generate interest and therefore greater involvement by the respondents, submersing them into the task at hand and therefore eliciting more useful responses. Frederickson (1986) recommends the use of structured interviews with industry experts in developing the instrument; these interviews play a critical role in achieving a rich and detailed understanding of the language and economic problems and operating realities of the particular industry. Without this intense industry knowledge, scenarios risk being interpreted as sterile and simplistic and thereby generate decreased respondent involvement (Frederickson 1986). Since the task of developing and validating vignettes is not an easy one, it is recommended that researchers make use of scenarios previously established in the literature if at all possible (Weber 1992; Wason et al. 2002). Several resources offering advice on a structured and comprehensive vignette design and validation process also exist (Frederickson 1986; Rungtusanatham et al. 2011).

While typically the decision scenario implies a written case description, they may also be portrayed "live" using audio and/or visual executions of the decision scenario (Ashill and Yavas 2006). Caro et al. (2010) suggest numerous advantages to using audio/visual decision scenarios, including: (1) more fully engaged research participants; (2) increased believability and reduced framing effects; and (3) increased interactivity and opportunities for the respondent to seek additional information, similar to approaches employed in real-world decision-making tasks. Of course, the authors are also quick to point out the challenges associated with audio/visual implementation, particularly in regard to the technological require-ments. Cross-platform functionality is a concern, as is the ubiquity of the software employed. Finally, some populations may lack the skills or confidence required to use these more complex instruments, and so this must be taken into consideration during design, as well.

One critical decision involves the number of vignettes to employ in a study. This is affected by the number of critical variables and their levels. The research design can become quite unwieldy with even minimal variation; for example, while five dichotomous variables lead to a total of 32 vignettes (2^5), the addition of just two more critical variables increases the number of vignettes required to an astounding 128 (2^7). Certainly, while the research question should drive the number of variables to investigate, and thus the length and content of the vignettes, researchers must be mindful of the risks of using too few or too many vignettes. At one end of the spectrum, a constant variable value vignette (CVVV) may be used, in which the researcher administers an identical vignette to all the respon-dents in a study (Cavanagh and Fritzsche 1985). While easy to develop and administer, this technique hinders analytical potential.

Vignettes that are systematically varied to accommodate different levels of factors are called contrastive vignettes (Cavanagh and Fritzsche 1985). This results in a richer cadre of data to analyze, and is especially revealing in determining the effects of changes in combinations of variables. As illustrated by Alexander and Becker (1978, p. 95), "most people are not particularly insightful about the factors that enter their own judgment-making process", and this is likely particularly true where interactions between variables exists. Too many variations, however, may lead to respondent fatigue and information overload (Weber 1992). The solution to complex designs is to employ a fractional replication design. Appropriately executed, these designs minimize confounding effects.

Ultimately, researchers are interested in studying the attitudes, beliefs, perceptions, and norms of the target population (Ashill and Yavas 2006), and as such introduce response items asking the respondents how they would deal with or respond to the situation presented in the vignette. Response items are oftentimes closed ended, presenting the respondent with a menu of options from which to choose between. This multiple choice format bounds the solution possibilities available to the respondent, and is largely unrepresentative of how decision-makers must process problems in the real world (Randall and Gibson 1990). Alternatively, open-ended questioning may be used, but is difficult and time-consuming to code and also requires multiple researchers for achieving inter-rater reliability (Weber 1992). Despite these challenges, several researchers advocate the use of open-ended over closed-ended questioning (Finch 1987; Randall and Gibson 1990).

As with all methodologies, there are certain advantages and disadvantages to the vignette technique. Benefits to employing vignette research as opposed to direct-question-based research include: (1) greater realism; (2) use of standardized stimuli, which improves internal validity, measurement reliability, and is easily replicable; (3) enhanced construct validity through focus on specific features; (4) more cost effective and more quickly executed than field-based studies of decision-making processes; and (5) potential to reduce social desirability bias of respondents through the use of third-person framing of vignettes (Wason et al. 2002).

A key limitation with vignette research is that unlike empirical field research, this technique only assesses "*facsimiles* of real situations and the subjects' responses to the scenarios demonstrate *intended* reasoning, decisions, or behavior" (Weber 1992, p. 147, italics in original). Vignettes should be pretested for representativeness, and posttests administered to assure respondent understanding (Weber 1992). It is not uncommon for researchers to leverage the captive audiences available within their classrooms. However, it is critical that the vignette matches the population for understanding, familiarity, and generalizability of the results (Weber 1992). Since vignette studies may be conducted by mail or Internet, it is not unreasonable to expect targeting of the appropriate managerial populations in these studies. Some validity issues in vignettes are common across all forms of experimentation, and are elaborated upon in the following section.

3.2 Process Simulations

An alternative to classical vignette studies is the use of multiperiod process simulations. These may involve either physical or computerized tasks, and may be implemented either as facsimiles of reality or, in the rarer case, to coincide with a natural experiment observed in situ. A fundamental distinction between process simulation studies and static vignettes approaches is typically the ability to observe actions taken and decisions made by subjects over a series of periods, representing multiple occurrences of stimulus–response. In that respect, a host of objective measures are often available for collection across an examination period including the time for individuals to complete tasks, or switch between tasks, the total number of errors made and more generally deviation from what might be defined as rational or optimal decision-making (c.f. Bearden et al. 2008; Schultz et al. 1998). More recently, however, still more intrepid attempts have been made to capture objective biometric data coinciding in time with specific stimuli presentations and response decisions (Seawright and Sampson 2007; Bendoly 2011). Multiple observations make possible the testing of research questions that involve event-driven or auto-corollary change (e.g. learning, reciprocity, etc.) and can also be useful in reducing the error in estimates of overall subject characteristics (e.g. risk aversion, ego-centrism, etc.).

3.2.1 Laboratory Simulations

Because natural experiments typically bring along a number of uncontrollable externalities, controlled laboratory experiments for process simulations have been more popularly used in recent years (perhaps, the most notable example being process simulations involving the 'Beer Game', see Sterman 1989; Croson and Donohue 2003, 2006). To be sure, the nature of the operations management context examined in such laboratory studies tends to be somewhat if not highly stylized. As is the nature of most modeling, the simulations designed to provide stimuli and response options to subjects are limited—constrained by tractability requirements and a general interest in focusing on a small subset of behavioral phenomena. No one would argue that the Beer Game, for example, is a realistic depiction of today's modern supply chain; however, the design of the game continues to be useful in studying very specific behavioral reactions (not to mention a simple way of demonstrating such reactions to a class of students).

Laboratory experiments are useful for several different purposes of research, including investigating theory, examining anomalies, and evaluating new policy or process (Roth 1986; Croson 2002; Croson and Gachter 2010). Each of these purposes requires special considerations in the design and execution of a laboratory experiment. Experiments can address theory by applying direct tests of theory, conducting comparisons of competing theories, and assessing the parameters of a particular theory or boundaries at which a theory breaks down. In experiments addressing theory, issues of internal validity are of utmost importance. In other words, to accurately test a theory, it is critical that the research capture exactly and fully all the assumptions embodied in the theory. Similarly, internal validity plays an important role in the investigation of anomalies, an "observed regularity that is not consistent with or predicted by current models" (Croson 2002, p. 930). Experiments designed to investigate anomalies seek to determine why the anomaly is observed and under what conditions it materializes. Finally, laboratory experiments are useful for demonstrating the parameters and unintended consequences of new policies or processes. Unlike the previous two experiment types, policy experiments rely on a high degree of external validity. It is extremely important in these cases to model the environment (the people, the context, etc.) as closely as possible, so that findings from the laboratory can predict with greater confidence what will happen when the policy or process is actually implemented in the real world.

Despite these important differences, researchers conducting a laboratory experiment can expect to follow a fairly general and predictable set of overarching steps. These stages are: (1) the experimental design; (2) subject pool selection; (3) implementation, and (4) compensation (Croson 2002). Within these steps, we will highlight a few differences born out of the foundational discipline that an experiment is based upon. These differences primarily arise depending on whether the experiment is grounded in theories of psychology or economics. Croson (2005) identifies several points of divergence between the common experimentation practices within psychology and economics; of those, the issues of incentives, context, subject pools, and deception are most applicable and are integrated into our discussion below.

The first step, experimental design, includes determining the number of treatments to be incorporated into the effort. A careful balance must be struck between introducing a sufficient number of treatments for the research to be interesting and limiting the number of treatments so as to not create an unwieldy research endeavor. Three to six treatments are recommended (Croson 2002). In addition, it must be decided whether the effort will adhere to a between-subjects design, wherein each participant receives just one treatment, or a within-subjects design, wherein participants receive multiple treatments. It is not a trivial decision, as within-subject designs do allow the researcher multiple observations per participant, but the participant may benefit from some learning of the task as he or she progresses through multiple rounds of the experiment. In economics experiments, generally it is advisable to keep the experiment void of context. This mimics the theory being tested, and also serves to reduce variance and minimize bias that context tends to introduce. The exception to this is experiments that test new policies and processes, where external validity—and thus context—is critical. Finally, in this stage, aspects of the experiment should be tested on individuals not otherwise associated with the research. For example, the instructions provided at the outset of the experiment need to be simple yet informative enough for the subject to fully understand the task at hand, and the software (if used) needs to be easy to use and provide feedback in a way that is meaningful to the participant. Pre-testing and pilot tests of the experiment can catch any potential glitches before they become overly problematic.

The next consideration in controlled human experiments involves which humans to recruit to the subject pool. This, too, is a nontrivial decision, and one that has led to much discussion within the academic community, particularly regarding when the use of students is acceptable. The first consideration should likely be the type of experiment being conducted (Croson 2010; Stevens 2011).

For example, most theories make no assumptions about which groups of people the theory applies to and does not apply to—they are general theories meant to apply broadly. Therefore, when testing theory it is generally considered acceptable to employ students as subjects. Alternatively, as the context of the experiment becomes more relevant and complex, the population becomes more specific. In this case, then, students may or may not be appropriate subjects. The problem again is largely one of validity—where internal validity is the primary consideration, student samples are generally suitable; where external validity takes precedence, students are likely not a suitable proxy (unless students are the object of study) (Stevens 2011). Other considerations include availability of subjects for the study (especially where attendance in a laboratory is required), and payment of subjects (where a professional's time is worth more than a student's), but these aspects should not be drivers of the decision regarding subject pool.

The third stage in the laboratory experimentation process is implementation. A critical goal in any experimental situation is for the experimenter to reduce as much as possible any "noise" in the procedures employed. Proper procedure, in accordance with McGuigan (1978), is to limit the number of randomly occurring extraneous variables, such that error variances are reduced. It is important when scheduling subjects to the laboratory to make use of random assignments so as to avoid potential confounding effects. As multiple sessions will likely be adminis-tered, it is advised that all participants in all groups be treated identically. The importance of using the same words, and even the same intonations, is not to be overlooked (McGuigan 1978). It is recommended that instructions at the outset of the experiment be read aloud from a script, and it may be beneficial to play a tape recording of the instructions to further minimize variation. Frequently, researchers find it useful to test subjects on their comprehension of the task, particularly if the decisions are relatively complex. This is ok, but tests should be carefully con-structed so as to avoid introducing any demand effects, or premonitions as to the purpose of the experiment (Croson 2002). As discussed previously regarding vignettes, it is possible that process simulations of this nature be administered not in a physical laboratory, but over the Internet instead. The same considerations must be taken into account in this case, particularly regarding the instructions, cross-platform functionality, and software.

The final step to the experiment process is the compensation of subjects. Most experiments offer some form of compensation to participants as incentive to get them into the laboratory. This incentive may take the form of extra course credit, a flat fee, or an earnings-based fee (with or without an additional show-up fee). The benefit of the latter incentive is that it motivates participants to perform well on the task assigned to them, versus if a flat fee were offered and participants could lessen their level of effort as it would have no effect on their personal outcome. This is an important consideration in economics-based experiments as compared to psychol-ogy-based ones. Economics experiments rely on induced valuation in their payoff schemes (Smith 1976). This is due to the fact that economic experiments evaluate the decisions or choices people make, and the underlying theories assume specific payoffs that the experiments must account for to understand actual decision-making.

It is an issue of internal validity of the experiment. In experimental psychology, typically the interest is in evaluating thought processes or attitudes, and so a flat fee incentive is more commonplace and generally acceptable.

The topic of deception is a consideration throughout the entire process outlined above. It is a fairly stringent rule in economics experiments that researchers not deceive their subjects on any aspect of the experimental design, subjects' roles, their counterparts in the experiment, or their payoffs. The rationale is that participant behavior is affected by deception, and that subjects who have been deceived before lack trust in future experiments they take place in; in other words there may be "reputational spillover effects" that compromise the nature of experimental studies (Hertwig and Ortmann 2001, p. 397). In psychology experiments, the forbearance on deception is not an issue, and some researchers acknowledge that it is "often a methodological necessity" (Kimmel 1996, p. 68) "to examine situations which would not occur naturally, for example, how individuals respond to low ultimatum offers" (Croson 2005, p. 140). Regarding this and other differences between experimental psychology and economics, however, Croson points out that "there are no right and wrong answers" (2005: p. 145). Researchers must look to the purpose of their study and conduct it in a way that makes sense, while being respectful of their subjects and mindful of their colleagues' work, as well. While many of the current laboratory studies in OM are informed by the practices of experimental economics (testing normative theories and the decisions people make related to those theories), there have been numerous calls for research investigating thought processes and social applications (Gino and Pisano 2008; Donohue and Siemsen 2011; Eckerd and Bendoly 2011). It will likely follow that the work, and therefore the practices, established within experimental psychological may be more often integrated into OM laboratory studies in the future.

3.2.2 Natural Experiments

Natural experiments, sometimes referred to as industrial or field experiments, are those in which real workers are observed performing their actual job duties in real time (Bendoly et al. 2006). Experiments of this type are useful for investigating phenomena of a socio-technical nature; in other words, interactions of changes in the technologies and processes being employed with the social systems supporting them (Huber and Brown 1991). Where most laboratory experiments strive to achieve high levels of internal validity, natural experiments—like experiments that testbed policies and processes—aim to achieve high levels of external validity. This is evident in that natural experiments take place in the field using actual workers and processes. This in many ways makes up for the loss of realism associated with laboratory experiments, and retains the multi-period phenomenological observation properties that vignettes lack. Another advantage to natural experiments is that the participants may not even be aware of their participation in a research effort, which can reduce bias (Greenberg and Tomlinson 2004).

Taking an experiment to the field is not a simple task, however. With a field experiment, the researcher is attempting to find an appropriate balance between control and naturalism, but by definition, moving out of the laboratory results in a loss of much of the control afforded by that "clean" environment (Greenberg and Tomlinson 2004). Various additional weaknesses and criticisms relating to field experiments have been identified. Primarily, the lack of control leads to the potential for numerous and unidentified confounding variables (Schwenk 1982). Moreover, the variables of interest may be of low quality and/or multidimensional, and as such, the task of parsing out individual relationships is hindered (Greenberg and Tomlinson 2004). This effect is evident in Greenberg (2002), who observed the consequences of implementing an ethics code in office environment versus the absence of such a code in another. While the overall benefit of having an ethics code could be inferred, the actual pieces of the code that were most effective could not be identified through the particular design executed. This example illustrates also the quasi-experimental nature of many field experiments. Specifically, random assignments to treatments are not always possible as the groups or individuals selected are determined by someone other than the researcher, for example, the host company (Greenberg et al. 1999). Finally, if multiple different treatments are run within the same organization, it is possible that subjects in different treatments will communicate with one another and potentially lead to adverse enlightenment effects (Gergen 1973).

The challenges associated with natural experiments should be viewed as just that, however—challenges to be overcome through thoughtful experimental design and rigorous adherence to procedures. For example, design of natural experiments can be enhanced through the use of multiple comparison groups, and/or multiple treatment groups that take place in different settings or at different levels of intensity (Meyer 1995). The potential insights to be gained through the execution of natural experiments make the effort of venturing out into the field worthwhile, as numerous calls for research employing this methodological tract have demonstrated (Bendoly et al. 2006; Fisher 2007; Craighead and Meredith 2008). In likely the best case scenarios, mixed methods research efforts combining the benefits of not only laboratory and field studies, but also those of other methodologies, will offer the biggest rewards (Schwenk 1982; Meredith 1998; Gupta et al. 2009). We turn to a look at mixed experimental studies next.

3.3 Mixed Experimental Studies

Having outlined some of the common approaches used in behavioral experimentation, it is worth emphasizing that there are opportunities in which the joint use of more than one of these approaches may prove useful. For example, although vignettes have served in numerous studies as stand-alone methods, they can and have also been employed in mixed studies as a backdrop to process simulations. In such applications, vignettes can serve to prime individuals for more ideal experimental responses; hence, permitting greater clarity in the analysis of research questions. Alternately, it is possible to imagine recorded sessions of individuals engaged in a behavioral experiment to serve as a component of a fairly rich vignette. In such use, researchers would speculate on the specific behavioral responses of subjects viewing alternate dynamics depicted by distinct recorded sessions (e.g. Which team appeared more cohesive/efficient/effective? What were their greatest strengths/weaknesses? Who would you outsource X project/process to given the choice?). Such a mixed method study could provide considerable insights into higher level operations management, project management, or even COO decisions, and latent priorities.

4 Validity: Interpretation of Design and Response

Measurement is a process that involves linking underlying theoretical concepts to empirically grounded indicators (Carmines and Zeller 1979). The validity of a measurement is one indication that the linkage between the (empirically) observable and the (theoretical) unobservable that has been proposed is a strong, high quality linkage from which useful inferences can be drawn. A measurement is valid if it succeeds in measuring what it is intended to measure, and measures nothing else. Success in validation is a matter of degree, however, and the process of establishing validity is not an easy one (Carmines and Zeller 1979; Flynn et al. 1990). Despite the difficulties associated with establishing validity, Nunnally and Bertnstein (1994) identify the issue of validity as the most important in psychometrics. Flynn et al. (1990) stress the importance of considering issues of validity during all stages of an empirical research effort, in order to enhance generalizability of the study's results. The importance of validity testing, therefore, is of critical importance in experimental research. As Bachrach and Bendoly (2011) point out, rigorous adherence to these most basic tenets of experimental research helps to ensure findings that are relevant and reliable, and thus make substantive contribution to our field.

4.1 Validity's Role

Validity generally speaks of the appropriateness or meaningfulness of measurements (Rosenthal and Rosnow 1991). In experimental research, four assessments of validity are typically recognized: external, internal, construct, and conclusion (Cook and Campbell 1979). External validity deals with the generalizability of conclusions drawn from the research to situations beyond the laboratory and involving different people, places, and time periods. Internal validity, only relevant in studies of causality, means the observed changes can be attributed to the independent variables intended, and not to other possible causes. These alternative explanations, referred to as confounding variables, must be identified and controlled for in the design of the experiment. Internal validity is generally less relevant in observational or descriptive studies. In pursuing internal validity, it is suggested that cause and effect can be established via three criteria: (1) temporal precedence; (2) covariation of cause and effect; and (3) a lack of plausible alternative explanations (a research design issue).

Construct validity represents the degree that the actual (or operationalized) construct reflects the ideal. It reflects the degree to which inferences can legitimately be made. According to Trochim and Donnelly (2007), there are multiple measurement-related validity terms demonstrating different aspects of construct validity:

- *Face validity*—this is essentially a subjective judgment assessing the quality of a measure; use of experts to make the judgment is recommended.
- *Content validity*—is a congruence of the operationalization and the relevant content domain.
- *Predictive validity*—degree to which the operationalization is able to predict something it theoretically should be able to predict; test this through correlations.
- *Concurrent validity*—degree to which the operationalization is able to distinguish between groups that it theoretically should be able to distinguish between.
- *Convergent validity*—degree to which the operationalization is similar to other operationalizations to which it theoretically should be similar to; assess this by high correlations between the operationalizations.
- *Discriminant validity*—degree to which the operationalization is not similar to other operationalizations to which it theoretically should be dissimilar to; assess with low correlations between the operationalizations.

Construct validity can be assessed through use of a nomological network, developed by Cronbach and Meehl (1955). It is essentially a "philosophical foundation", or a visual representation of the constructs employed in a study and how those constructs interrelate. A form of nomological network is used in the design of structural equation models, which presents a mathematically rigorous way to assess constructs and construct relationships.

Finally, conclusion validity regards the soundness of inferences drawn from the data analysis. In other words, to what degree are the conclusions drawn reasonable? Tireless efforts during the design of experiments help achieve validity in experiments, but as is often true, even our best-laid plans may go awry. Moreover, achieving validity is not an all-or-none proposition; for example, we may have demonstrated internal validity but be lacking construct validity.

4.2 Threats to Validity and Minimizing Threats by Design

Threats to validity may be ameliorated through a number of means (c.f. Podsakoff et al. 2003). Random selection and replication, for example, can be used to ensure that subject assignment to treatments is free of structural biases. Ensuring a testing environment where subjects feel safe and able to focus on the task at hand undisturbed by potential external sources of noise is also a crucial element of experimental context. Piloting the task and experimental environment of course is a fundamental mechanism for pretesting the effectiveness of the design and its prospective validity. Exit interviews conducted at the pilot stage, as well as the main study, can help ensure where failures in the design exist or evolve. The content of such interviews can be absolutely invaluable in identifying unanticipated problems with the task or experimental design overall. In addition to these generally applicable guidelines, means for minimizing the specific threats of external, internal, construct, and conclusion validity are available.

Threats to external validity include the people, place, and time about which the researcher is making a generalization. Proximal similarity models serve to map out gradients of similarity, and thus generalizability of the results of a study to different groups (Campbell 1986). Additionally, replications of a study across different people, contexts, and times enhance external validity. Threats to internal validity include:

- Single-group threats, which may be remedied through the use of a control group.
- Multiple group threats, which are evident when the groups are not comparable prior to the implementation of the study or treatment. This is also referred to as selection bias. Use of randomization serves to prevent this threat.
- Social threats, which include pre-existing knowledge of the experiment by the participants.

Threats to construct validity (as developed from Cook and Campbell 1979) are composed of both design threats and social threats. Design threats include the following:

- An inadequate preoperational explication of constructs. The remedy involves the comprehensive literature reviews and expert (albeit subjective) assessment of operationalizations. Clarity of concept definition and intent, as well as the avoidance of confusing descriptions/instructions can safeguard against misinterpretation of experimental tasks, objectives, and rewards.
- Mono-operation bias, which may be remedied through the use of multiple replications with respect to people, place, and time.
- Mono-methods bias, which is remedied through the application of multiple methods.
- Interaction of treatments, which can be planned for and identified through the use of a control group.
- Interaction of testing and treatment, which may be prevented by introducing a control group, or through use of a Solomon 4-group design (Campbell and Stanley 1963).
- Restricted generalizability across constructs (on other words, unintended consequences). The best remedy for this threat is to anticipate and measure all potential outcomes.

- Confounding constructs and levels of constructs which calls for a comprehensive examination of the *ranges* of effectiveness.

  Social threats include the following:

- Hypothesis guessing by participants. This implies that the participants behave in a way that they believe the researcher wants them to. The appropriate remedy is the use of control group. Alternatively, the researcher may also attempt to hide real purpose of the study from participants.
- Evaluation apprehension, where the participant is uncertain of task. Remedy this by task training.
- Experimenter expectancies, where the researchers clues the participant (knowingly or unknowingly). Carefully crafting scripts and adhering to them is useful for overcoming this threat.

Finally, threats to conclusion validity can be classified into Type I and Type II causes. Type I threats mean the researcher has identified a relationship when in fact there is none. Often this occurs when multiple analyses are conducted and the error rate is not sufficiently adjusted to account for them. Type II threats involve failure to find an existing relationship. This may occur due to a low reliability of measure, poor reliability of treatment implementation, random irrelevancies, random heterogeneity of participants, or low statistical power. This may also be due to a violation of the assumptions of the particular statistical tests employed.

4.3 Validity Testing

In order to demonstrate the validity of the experiment carried out, checks to the clarity of controlled treatments imposed on the design must be made. Without checks to validate the roles of specific treatments, the conclusions drawn with respect to the impact of the treatment classes acting on key dependent variables may quickly become suspect. As a result, the credibility of behavioral experiments hinge on such validation, particularly when results are intended to be extrapolated toward practical application or subsequent theory development.

At least three classifications of treatment checks can provide meaningful support for researchers. Those checks that serve to assess the ability of the treatment to characterize differing levels of an intended construct (i.e. *manipulation checks*) focus on the convergent validity of the treatment. Manipulation checks are often best conducted through the use of well-developed or established multi-item scales indicative of each treatment, and the collection of subject responses to these items following soon after the treatment application. Comparative statistics (e.g. *t*-Tests, ANOVA, etc.) are often used to test delineations of treatment levels, and thus support convergent validity.

Other checks, focused on discriminant validity, serve to ensure that individual treatments do not confound other theoretically 'independent' issues of interest. These secondary checks are often referred to as *confounding checks* (Wetzel 1977), and are often tested through comparative statistics as well—in this case testing whether the treatment levels inadvertently impact perceptions of other supposedly independently controlled issues. Both confounding and manipulation checks are particularly helpful in the "pre-test" and pilot phases of studies to ensure validity of the main experiment, though they should be included as part of the main experimental analysis as well.

Hawthorne checks (Mayo 1949; Adair 1984; Parsons 1992) against extraneous perceptual effects of treatments constitute a third validity test. Such checks are often conducted using supplemental measures not viewed as critical to the research questions studied but thought to be nevertheless related to the context studied. Successful results of such checks should suggest no impacts from any of the treatments on supplemental measures otherwise assumed to remain independent of the study. In this example, such supplemental measures might include customer perceptions of the convenience of the bank's "location". Perceptions of the availability of seating (or parking in a more realistic setting) would not be a reasonable measure for use in Hawthorne checks since line length and its relationship to staffing and throughput can reasonably be viewed as intertwined with such measures. Therefore, successful validity checks of this nature require both an appropriate selection of supplemental measures as well as results that suggest they are not impacted by the design's treatments. If impacts are found, then the focus and isolation of the treatments can be called into question—and thus the clarity of the relationships analyzed.

While data used in rigorous application of treatment checks tend to be collected through numeric scales or objective observations as part of the experiment, or pre- and post-experimental surveys, unstructured exit interviews can also prove informative in the matter. If unstructured interviews suggest a blurring of concepts in the mind of the subjects or a general misunderstanding of specific treatment levels, validity can be called into serious question if not rejected outright. Summary analysis of the content of such interviews should accompany claims of treatment validity whenever available. Increasingly, it is likely that such thoroughness of evidence will be expected of researchers in this area.

It should be emphasized that to date, the vast majority of OM behavioral studies have failed to adequately provide for the above checks.

5 Lessons for Operations Management and Future Questions

Having outlined various issues and tactics associated with the use of behavioral experiments in OM research, it is worth closing with some of the more interesting general findings that emerge from behavioral experiments in OM research in the last decade. These phenomena represent some of the most interesting ideas resulting from experimentation and provide ample ground for future experimental investigation.

5.1 System Comprehension Effects

Perceptions of system dynamics based on limited information cues influence tactical decision-making in multiple OM settings. One of those contexts strongly effected seems to be that of project management. Bendoly et al. (2010a, b) and Bendoly and Swink (2007) demonstrate that perceptions task complexity as well as the staff-sharing behavior of other project managers impacts the tendency of individuals to seek out globally optimal tactics in the management of their own projects. The impact of systems perceptions relating to workload is also clearly manifested in physiological displays of stress and awareness, which seem to be related to the ability of revenue managers (for example) to gain insights from operational decision support tools (Bendoly 2011).

5.2 Nonmonotonic Behavioral Dependency

Workload also seems to have a significant effect on the nature of behavior among both operations workers and managers (Schultz et al. 1998, 1999, 2003). Importantly, however, experimental and other empirical studies seem to suggest that a strong nonmonotonicity (inverted-U) exists between workload levels and response (Bendoly and Prietula 2008; Bendoly and Hur 2007; Choo et al. 2007). Where specifically in the realm of work for a given task this point of inversion takes place is not clear, it makes the effect highly context-specific. This in turn makes prescriptions of workload management much more difficult to appropriately develop, and casts a great deal of doubt on existing prescriptions that have viewed workload as having either a monotonic or worse still a noneffect on worker behavior.

5.3 Unstable Behavioral Processes

There has been a long tradition appreciating the effects of learning in operations management. Unfortunately, the microfoundations of learning (learning at the individual or group level) have been given little attention in OM studies. This has tended to be problematic for OM prescriptions that do not sufficiently account for changing dynamics over time. Bendoly and Cotteleer (2008) demonstrated that learning how to misuse large implemented technologies for example strongly relate to losses in initial gains made possible by these systems (an effect they attribute to learning as well as something they refer to as resonant dissonance). Bendoly and Prietula (2008) also suggest that the inflection point of a nonmonotonic workload-performance curve greatly depends on the extent to which individuals are acquainted with a process (i.e. extent of workload and hence the optimal workload level are dependent on task familiarity which is a nonstatic concept in learning contexts).

Now that these and other behavioral phenomena have proven salient to established OM research contexts, it is absolutely incumbent on new research in these contexts to take such phenomena into account when addressing new research questions. Furthermore, it should be of interest to OM researchers to investigate past research that failed to account for such issues, and as a consequence failed to see application in practice. It may be that many normative models existing in the literature are only steps away from real practical impact, save for their lack of incorporation of the phenomena outlined here and illustrated by other OM behavioral experiments. With a willingness to now rigorously consider such issues, the future of OM research becomes a much broader and potentially more influential domain.

References

J.G. Adair, The Hawthorne effect: a reconsideration of the methodological artifact. J. Appl. Psychol. **69**(2), 334–345 (1984)

C.S. Alexander, H.J. Becker, The use of vignettes in survey research. Public Opin. Q. **42**(1), 93–104 (1978)

N.J. Ashill, U. Yavas, Vignette development: an exposition and illustration. Innov. Mark. **2**(1), 28–36 (2006)

D. Bachrach, E. Bendoly. Rigor in behavior experiments: a basic primer for SCM researchers. J. Supply. Chain. Manag. **47**(3), 5–8 (2011)

N. Bearden (2010), http://en.wikipedia.org/wiki/Behavioral_operations_research

J.N. Bearden, R.O. Murphy, A. Rapoport, Decision biases in revenue management: some behavioral evidence. Manuf. Serv. Oper. Manag. **10**(4), 625–636 (2008)

E. Bendoly, Linking task conditions to physiology and judgment errors in RM systems. Prod. Oper. Manag. **20**(6), 860–876, (2011)

E. Bendoly, M.J. Cotteleer, Understanding behavioral sources of process variation following enterprise system deployment. J. Oper. Manag. **26**(1), 23–44 (2008)

E. Bendoly, R. Croson, P. Goncalves, K. Schultz, Bodies of knowledge for research in behavioral operation. Prod. Oper. Manag. **19**(5), 434–452 (2010a)

E. Bendoly, K. Donohue, K. Schultz, Behavior in operations management: assessing recent findings and revisiting old assumptions. J. Oper. Manag. **24**(6), 737–752 (2006)

E. Bendoly, D. Hur, Bipolarity in reactions to operational 'constraints': OM bugs under an OB lens. J. Oper. Manag. **26**(1), 1–13 (2007)

E. Bendoly, P. Perry-Smith, D. Bachrach, The perception of difficulty in project-work planning and its impact on resource sharing. J. Oper. Manag. **28**(5), 385–397 (2010b)

E. Bendoly, M. Prietula, In 'The Zone': the role of evolving skill and transitional workload on motivation and realized performance in operational tasks. Inter. J. Oper. Prod. Manag. **28**(12), 1130–1152 (2008)

E. Bendoly, M. Swink, Moderating effects of information access on project management behavior, performance and perceptions. J. Oper. Manag. **25**(3), 604–622 (2007)

D.T. Campbell, Relabeling internal and external validity for applied social scientists. New Dir. Program Eval. **31**, 67–77 (1986)

D.T. Campbell, J.C. Stanley, *Experimental and Quasi-Experimental Designs for Research* (Houghton Mifflin Company, Boston, 1963)

E.G. Carmines, R.A. Zeller, *Reliability and Validity Assessment*. Sage University paper series on quantitative applications in the social sciences, 07-017, (Sage, Newbury Park, 1979)

F.G. Caro, T. Ho, D. McFadden, A.S. Gottlieb, C. Yee, T. Chan, J. Winter, Using the internet to administer more realistic vignette experiments. Soc. Sci. Comput. Rev. Published online 12 January 2011, **30**(2), 184–201 (2010)

G.F. Cavanagh, D.J. Fritzsche, Using Vignettes in Business Ethics Research. In: ed. by L.E. Preston. Research in Corporate Social Performance and Policy, vol 7 (JAI Press, Greenwich, 1985), pp 279–293

A.S. Choo, K.W. Linderman, R.G. Schroeder, Method and context perspectives on learning and knowledge creation in quality management. J. Oper. Manag. **25**(4), 918–946 (2007)

T. Cook, D. Campbell, *Quasi-Experimentation: Design and Analysis for Field Settings* (Rand McNally, Chicago, 1979)

C.W. Craighead, J. Meredith, Operations management research: evolution and alternative future paths. Int. J. Oper. Prod. Manag. **28**(8), 710–726 (2008)

L.J. Cronbach, P.E. Meehl, Construct validity in psychological tests. Psychol. Bull. **52**(4), 281–302 (1955)

R. Croson, Why and how to experiment: methodologies from experimental economics. Univ. Ill. Law Rev. **2002**(4), 921–945 (2002)

R. Croson, The method of experimental economics. Int. Negot. **10**(1), 131–148 (2005)

R. Croson, The Use of students as participants in experimental research. behavioral dynamics in operations management whitepaper (2010), www.ombehavior.com

R. Croson, K. Donohue, Impact of POS data sharing on supply chain management: an experiment. Prod. Oper. Manag. **12**(1), 1–11 (2003)

R. Croson, K. Donohue, Behavioral causes of the bullwhip effect and the observed value of inventory information. Manag. Sci. **52**(3), 323–336 (2006)

R. Croson, S. Gachter, The science of experimental economics. J. Econ. Behav. Organ. **73**, 122–131 (2010)

K. Donohue, E. Siemsen, *Behavioral Operations: Applications in Supply Chain Management*. Wiley Encyclopedia of Operations Research and Management Science (2011)

S. Eckerd, E. Bendoly, (2010), http://www.scholarpedia.org/article/Behavioral_Operations

S. Eckerd, E. Bendoly, Introduction to the discussion forum on using experiments in supply chain management research. J. Supply Chain Manag. **47**(3),3–4 (2011)

J. Finch, The vignette study in survey research. Sociology **21**(1), 105–114 (1987)

M. Fisher, Strengthening the empirical base of operations management. Manuf. Serv. Oper. Manag. **9**(4), 368–382 (2007)

B.B. Flynn, S. Sakakibara, R.G. Schroeder, K.A. Bates, E.J. Flynn, Empirical research methods in operations management. J. Oper. Manag. **9**(2), 250–284 (1990)

J.W. Frederickson, An exploratory approach to measuring perceptions of strategic decision process constructs. Strateg. Manag. J. **7**(5), 473–483 (1986)

K.J. Gergen, Social psychology as history. J. Pers. Soc. Psychol. **26**, 309–320 (1973)

F. Gino, G. Pisano, Toward a theory of behavioral operations. Manuf. Serv. Oper. Manag. **10**(4), 676–691 (2008)

D. Greenberg, M. Shroder, M. Onstott, The social experiment market. J. Econ. Perspect. **13**(3), 157–172 (1999)

J. Greenberg, Who stole the money, and when? Individual and situational determinants of employee theft. Organ. Behav. Hum. Decis. Process. **89**, 985–1003 (2002)

J. Greenberg, E.C. Tomlinson, Situated experiments in organizations: transplanting the lab to the field. J. Manag. **30**(5), 703–724 (2004)

S. Gupta, R. Verma, L. Victorino, Empirical research published in production and operations management (1992–2005): trends and future research directions. Prod. Oper. Manag. **15**(3), 432–448 (2009)

R. Hertwig, A. Ortmann, Experimental practices in economics: a methodological challenge for psychologists? Behav. Brain Sci. **24**, 383–451 (2001)

V.L. Huber, K.A. Brown, Human resource issues in cellular manufacturing: a socio-technical analysis. J. Oper. Manag. **10**(1), 138–159 (1991)

J. H. Kagel, A. E. Roth, *The Handbook of Experimental Economics* (Princeton University Press, New Jersey, 1995)

D. Kahneman, P. Slovic, A. Tversky, *Judgment Under Uncertainty: Heuristics and Biases* (Cambridge University Press, New York, 1982)

A.J. Kimmel, *Ethical Issues in Behavioral Research: A Survey* (Blackwell Publishing, Cambridge, 1996)

C.H. Loch, Y. Wu, Behavioral operations management. Found. Trends Technol., Inf. Oper. Manag. **1**(3), 121–232 (2007)

C.H. Loch, Y. Wu, Social preferences and supply chain performance: an experimental study. Manage. Sci. **54**(11), 1835–1849 (2008)

S.P. Mantel, M.V. Tatikonda, Y. Liao, A behavioral study of supply manager decision-making: factors influencing make versus buy evaluation. J. Oper. Manag. **24**(6), 822–838 (2006)

E. Mayo, Hawthorne and the Western Electric Company, *The Social Problems of an Industrial Civilisation*, (Routledge, London, 1949)

F.J. McGuigan, *Experimental Psychology: A Methodological Approach*, 3rd edn. (Prentice-Hall, Englewood Cliffs, 1978)

J. Meredith, Building operations management theory through case and field research. J. Oper. Manag. **16**, 441–454 (1998)

B.D. Meyer, Natural and quasi-experiments in economics. J. Bus. Econ. Stat. **13**(2), 151–161 (1995)

J.C. Nunnally, I.H. Bernstein, *Psychometric Theory* (McGraw-Hill, New York, 1994)

H.M. Parsons, Hawthorne: an early OBM experiment. J. Organ. Behav. Manag. **12**(1), 27–44 (1992)

P.M. Podsakoff, S.B. MacKenzie, J.Y. Lee, N.P. Podsakoff, Common method bias in behavioral research: a critical review of the literature and recommended remedies. J. Appl. Psychol. **88**(5), 879–903 (2003)

D.M. Randall, A.M. Gibson, Methodology in business ethics research: a review and critical assessment. J. Bus. Ethics **9**(6), 457–471 (1990)

R. Rosenthal, R. Rosnow, *Essentials of Behavioral Research: Methods and Data Analysis*, 2nd edn. (McGraw Hill, New York, 1991)

A. Roth, Laboratory experimentation in economics. Econ. Philos **2**(2), 245–273 (1986)

M. Rungtusanatham, C. Wallin, S. Eckerd, The vignette in a scenario-based role-playing experiment. J. Supply Chain Manag. **47**(3), 9–16 (2011)

K.L. Schultz, D.C. Juran, J.W. Boudreau, The effects of low inventory on the development of productivity norms. Manage. Sci. **45**(12), 1664–1678 (1999)

K.L. Schultz, D.C. Juran, J.W. Boudreau, J.O. McClain, L.J. Thomas, Modeling and worker motivation in JIT production systems. Manag. Sci. **44**(12), Part 1 of 2, 1595–1607 (1998)

K.L. Schultz, J.O. McClain, L.J. Thomas, Overcoming the dark side of worker flexibility. J. Oper. Manag. **21**(1), 81–92 (2003)

M. Schweitzer, G. Cachon, Decision bias in the newsvendor problem. Manage. Sci. **46**(3), 404–420 (2000)

C.R. Schwenk, Why sacrifice rigour for relevance? A proposal for combining laboratory and field research in strategic management. Strateg. Manag. J. **3**, 213–225 (1982)

K.K. Seawright, S.E. Sampson, A video method for empirically studying wait-perception bias. J. Oper. Manag. **25**(5), 1055–1066 (2007)

M.E.P. Seligman, S.F. Maier, Failure to escape traumatic shock. J. Exp. Psychol. **74**, 1–9 (1967)

V.L. Smith, Experimental economics—induced value theory. Am. Econ. Rev. **66**, 274–279 (1976)

J.D. Sterman, Modeling managerial behavior: misperceptions of feedback in a dynamic decision making experiment. Manage. Sci. **35**(3), 321–339 (1989)

C. K. Stevens, Questions to consider when selecting student samples. J. Supply Chain Manag. **47**(3), 19–21 (2011)

W. Trochim, J. Donnelly, The Research Methods Knowledge Base, 3rd edn. (Atomic Dog Publishing, Cincinnati, 2007)

K.D. Wason, M.J. Polonsky, M.R. Hyman, Designing vignette studies in marketing. Australas. Mark. J. **10**(3), 41–58 (2002)

J. Weber, Scenarios in business ethics research: review, critical assessment, and recommendations. Bus. Ethics Q. **2**(2), 137–160 (1992)

C.G. Wetzel, Manipulation checks: a reply to Kidd. Represent. Res. Soc. Psychol. **8**(2), 88–93 (1977)

Chapter 2
Complex Systems Methodologies for Behavioural Research in Operations Management: *NK* Fitness Landscape

Ilaria Giannoccaro

Abstract From a methodology point of view, most Behavioural Operations Management (BOM) studies have employed experiments. However, no reason, either theoretical or practical, exists to limit BOM to experimental research. In this chapter, I discuss my conviction that methodologies coming from complexity science have the proper characteristics to be successfully applied in BOM research, since real operating systems, such as processes, factories, organisations and supply chains, are complex adaptive systems (CASs) where human behaviour is the central driver. Moving from this assumption, I suggest applying complexity science in order to study operating systems in diverse OM contexts and I also propose research questions coherent with a complexity science approach. They concern how operating systems behave, adapt and show new orders in terms of processes, structures and performances. Then, I suggest the adoption of a simulation tool to study CASs to develop BOM models, i.e. *NK* fitness landscape. After reviewing the methodology and its main applications in organisational contexts, I propose how different OM contexts can be modelled and how behavioural factors both at an individual and at a population level might be operationalised through the methodology proposed. Finally, I formulate research questions that might be addressed by applying *NK* fitness landscape.

I. Giannoccaro (✉)
Department of Mechanics, Mathematics, and Management,
Polytechnic University of Bari, Viale Japigia 182 70126 Bari, Italy
e-mail: ilaria.giannoccaro@poliba.it

I. Giannoccaro (ed.), *Behavioral Issues in Operations Management*,
DOI: 10.1007/978-1-4471-4878-4_2, © Springer-Verlag London 2013

1 Introduction

People significantly affect how operating systems work and perform. Nevertheless, the traditional Operations Management (OM) literature has largely ignored the effect of human behaviour, or at most, has considered it a secondary effect. Traditional OM has incorporated the classical assumptions of neoclassical economics so that humans involved in the management of operating systems are modelled as fully rational decision makers, acting solely to optimise measures of economic value. Oversimplified models of goals, motivation, learning, creativity and of other aspects of human behaviour such as intelligence, risk attitude, overconfidence, conformism, rejection of ambiguity and complexity, have been largely applied (Simon 1955; Chopra et al. 2004; Bendoly et al. 2006; Loch and Wu 2007; Gino and Pisano 2008).

Therefore, it is neither surprising that there is abundant evidence of real operating systems behaving differently in practice from the theoretical predictions, nor that theoretical prescriptions fail to deliver their promised achievements.

Behavioural Operations Management (BOM) is a multi-disciplinary branch of OM, encompassing organisational behaviour, decision science and psychology, that explicitly studies the effects of human behaviour on the performances of operating systems and analyses strategies to improve them (Gino and Pisano 2008; Loch and Wu 2007). In particular, BOM explores deviations from rationality of the decision makers involved in the management of operating systems including factors affecting their behaviour (Siemsen 2009), with the aims firstly of providing a better understanding of how operating systems work and perform, and secondly of developing effective implications for the design, management and improvement of operating systems (Gino and Pisano 2008).

From a methodology point of view, most BOM studies have employed experiments. However, no reason, either theoretical or practical exists to limit BOM to experimental research. Loch and Wu (2007) observe to this respect that *"the equation of BOM with experiments seems narrower than the spirit to the attempt to expand OM to incorporate people issues"*. I entirely agree that there is no need to restrict BOM to one methodological approach, i.e. behavioural experiments.

Instead, it is my conviction that methodologies coming from complexity science have the proper characteristics to be successfully applied in BOM research, since real operating systems, such as processes, factories, organisations and supply chains, are complex adaptive systems (CASs) where human behaviour is the central driver. Indeed, in such systems there are a number of independent, multiple and heterogeneous human agents making decisions using heuristics and schemata, self-organising by interacting among each other and co-evolving with the rugged and dynamic environment in which they exist (Choi et al. 2001).

Complexity science offers suitable theories and methodological tools for studying the evolution of CASs. As such, it is well suited to studying the dynamics of operating systems, which is one of the main challenges in OM research (Pathak et al. 2007).

Furthermore, complexity science-based methodologies permit the development of OM models that can easily include behavioural factors. They are particularly suitable to modelling not only the properties of individuals, such as their personal abilities, attitudes and cognitive biases, but also the factors affecting social interactions and characterising groups and populations, which are new behavioural issues that should be included in OM models in accordance with the most recent trend in BOM research (Loch and Wu 2007).

The main advantage coming from the application of complexity science-based methodologies resides in the possibility of understanding how behavioural factors affect the working and evolution of operating systems, allowing them to emerge spontaneously, given the characteristics of the system and the behavioural factors included. Such methodologies allow the difficulty of predicting and understanding which individual agent strategies lead to a desired collective behaviour to be overcome. Moreover, compared to experimental methods, they are less expensive and more effective because experiments could not show all possible events (Loch and Wu 2007).

The aim of this chapter is thus twofold. First, I intend to develop a theoretical framework which classifies BOM research and can help to identify new BOM research directions. This framework is based on the traditional logic for incorporating behavioural factors into OM models, its novelty lies in the proposal to develop BOM models using complexity science methodologies. In this way, I identify opportunities coming from complexity science for doing BOM research and at the same time define the BOM research questions that require methodologies suitable to study CASs. Next, I intend to show how a complexity-science methodology, i.e. *NK* fitness landscape, should be applied to model operating systems and behavioural factors so as to study their dynamics in different OM contexts.

This chapter is organised as follows. First, I review the main BOM taxonomies so as to identify the most important behavioural factors to be included into OM models. Then, I discuss the theoretical framework I propose, which is based on complexity science. A discussion follows regarding the main characteristics of *NK* fitness landscape and how it can be applied to model different OM contexts and most of the behavioural factors. Finally, I describe future BOM research directions and present some research questions addressable through *NK* fitness landscape, formulated following the proposed approach in a few OM contexts.

2 BOM Research: A Review of the Taxonomies

Taxonomies of BOM research classify common assumptions made in the OM literature concerning human behaviour and provide a rationale for identifying possible research questions in many different OM contexts.

Bendoly et al. (2006) classify behavioural assumptions into three broad categories: Intentions, Actions and Reactions. A BOM researcher should question whether assumptions concerning the intentions, actions and reactions of decision makers are valid and whether they could affect the performance of a given system and in turn the model's recommendations regarding the system.

'*Intentions*' refer to the model's accuracy in reflecting the actual goals of the decision makers.

Common assumptions about decision makers' factor weighting, risk attitude and the existence of not monetary goals such as trust and justice belong to this category.

'*Actions*' refer to the rules or implied behaviour of human players in the model. The most important assumption in this category is to neglect individual differences. Differences exist in human work rate, cognitive limitations, motivations, ability to process feedback, communication methods and personal abilities.

'*Reactions*' refer to the human player's response to model parameter changes. They mainly concern the role of feedback and its impact on human behaviour and the implied rules regarding how decision makers learn, process feedback or are affected by environmental factors.

The taxonomy provided is critical in order to identify behavioural gaps in the different OM contexts such as product development, inventory and DC management, quality management, production and workflow management, procurement and strategic sourcing and supply chain management. The authors give examples of common assumptions and possible behavioural gaps in all the categories in the different OM contexts. Based on this, they provide research questions and potential research directions in BOM.

Similarly to Bendoly et al. (2006) and Gino and Pisano (2008) focus on the cognitive limits of individuals and on the consequent systematic biases that might alter their decisions. They offer a wider classification of the cognitive biases made by individuals in the decision making process. They are classified on the basis of the stage of the decision-making processes in which they occur: (1) the information acquisition stage, (2) the information processing stage, (3) the outcome stage and (4) the information feedback stage (Table 1).

Individual biases could affect human behaviour in many different OM contexts such as product development and R&D, project management, supply chains, forecasting, inventory management, services and management of IT, which are similar in that they involve the acquisition, processing and interpretation of information from different sources. In each setting OM specific biases may occur and affect performances. Here, BOM research is greatly needed.

Gino and Pisano (2008) also suggest that BOM research should follow two complementary directions, i.e. prescriptive and descriptive. The prescriptive direction means that studies are needed which incorporate behavioural factors into OM models. The aim of such research would be normative: to provide valid recommendations and prescriptions on the design, management and improvement of operating systems, thanks to the extension of simplistic behavioural assumptions through improved modelling of human behaviour. The descriptive direction means developing studies that may improve our understanding of the effects of the cognitive biases on system performances so as to provide strategies and interventions to enhance them.

Table 1 Individual biases at the decision-making stage (Gino and Pisano 2008)

Bias	Description
Acquisition of information	
Information avoidance	People's tendency to avoid information that might cause mental discomfort or dissonance
Confirmation bias	People's tendency to seek information consistent with their views or hypotheses
Availability heuristics	People's tendency to judge an event as likely or frequent depending on the ease of recalling and imagining it
Salient information	People's tendency to weigh more vivid information than abstract information
Illusory correlation	People's tendency to believe two variables co-vary when they do not
Procrastination	People tendency to defer actions or tasks to a later time
Processing of information	
Anchoring and adjustment heuristic	People's tendency to rely too heavily, or anchor, on one trait or piece of information when making decisions
Representativeness heuristics	People's tendency to assume commonality between objects of similar appearance
Law of small numbers	People's tendency to consider small samples as representative of the population from which they are drawn
Sunk costs fallacy	People's tendency to pay attention to information about costs that have already incurred
Planning fallacy	People's tendency to underestimate task-completion time
Inconsistency	People's inability to use consistent judgment strategy across a repetitive set of cases or events
Conservatism	People's failure to update their opinions or beliefs when they receive new information
Overconfidence	People's tendency to be more confident in their own behaviour, opinions, attributes and physical characteristics they ought to be
Outcome	
Wishful thinking	People's tendency to assume that because one wishes something to be true or false then it is actually true or false
Illusion of control	People's tendency to believe that they can control or at least influence outcomes that they have no influence over
Information received through feedback	
Fundamental attribution error	People's tendency to overemphasise dispositional or personality-based explanations for behaviour observed in others while underemphasising situational explanations
Hindsight bias	People's tendency to think of events that have occurred as more predictable than they in fact were before they took place
Misperception of feedback	People's tendency to misperceive dynamic environments that include multiple interacting feedback loop, time delays and nonlinearities

While Bendoly et al. (2006) and Gino and Pisano (2008) exclusively focus on individual cognitive biases, Loch and Wu (2007) extend this view. They start by observing that the aim of BOM is to "*bring people issues back into the discipline*", and add further insights to BOM research, recognising that BOM should encompass behavioural factors concerning people's motivation in social interactions and the influence of group dynamics.

Based on this, they identify three behavioural OM categories. The first category is concerned with individual decision-making biases due to cognitive limitations; the second category is referred to individual behaviour driven by social goals in the context of social interactions. This category includes emotional signals motivating human behaviour in social interactions, such as the value given to status, fairness in relationships and a positive social image. The third category concerns collective behaviours that emerge in groups such as culture, knowledge and skills resulting from interacting learning processes activated within a given population.

These novel factors to be incorporated in OM models are considered to be more important than the cognitive biases and path the way to the introduction of new research methodologies particularly suited to dealing with interactions and group dynamics.

Summarising, I suggest two taxonomic criteria for classifying BOM studies: (1) models including the characteristics of individuals, among which, in turn, can be distinguished cognitive biases (for a list see Table 1) and social features such as reputation, social status, fairness; (2) models including the properties of groups and populations such as culture, norms, knowledge and skills.

3 A Framework for BOM Research: Opportunities from Complexity Science

My framework for BOM research is based on the claim that operating systems are CASs (Choi et al. 2001; Surana et al. 2005; Pathak et al. 2007; Bozarth et al. 2009), in which human behaviour is central in determining the evolution trajectories and performances. Indeed, in all OM contexts there are a variety of individuals making decisions on many different aspects and interacting among each other at diverse levels, so that the system is able to spontaneously self-organise and co-evolve with the dynamic environment assuming a new ordered configuration and new properties.

Complexity science is the discipline devoted to the study of CASs (Holland 1995). It provides theories and tools to explain how CASs behave and evolve, making it suitable for adoption in OM study contexts. I therefore propose to resort to complexity science to build OM models.

In the following section, I give theoretical support to the proposed framework for BOM research. First, I discuss the complexity of the OM settings and, then, frame them as complex organisational systems. Then, I suggest BOM research directions by discussing what research questions are feasible through this theoretical approach.

3.1 The Complexity of OM Context

OM is a multi-disciplinary discipline that investigates the design and management of operating systems, i.e. those systems involved in the development, production and distribution of products and services in the hands of final consumers. Typical OM contexts are:

- Product development and R&D;
- Project management;
- Inventory management;
- Production and workflow management;
- Procurement and strategic sourcing;
- Supply chain management.

A wide body of literature has underlined that operating systems in any context are complex organisations (Choi et al. 2001; Surana et al. 2005; Pathak et al. 2007). Complex organisations are complex systems made up of a large number of parts (agents) that interact among each other in nonlinear ways (Simon 1962). Nonlinearity means that there is not a direct correlation between the size of the cause and the size of the corresponding effect. Nonlinearity implies difficulty in making predictions dependable. Variety is a further fundamental property of the agents in a complex system (Casti 1997).

All the dimensions of complexity are recognisable in OM contexts. For example, supply chains are made up of heterogeneous firms each accomplishing a phase of the production process and interacting together to deliver a product/ service to the final customer. Variety characterises supply chains because firms differ in organisational culture, size, location and technology. Supply chains also show nonlinear behaviour such as the bullwhip effect (Choi et al. 2001).

These sources of complexity make the design and management of operating systems a very hard task. The traditional approach to handling complexity in OM contexts is to try to reduce it, for example, employing strategies aimed at reducing the number of parts, their variety and the links among them. Conversely, I suggest employing complexity science from both the theoretical and the methodological point of view.

3.1.1 Framing Operating Systems as CASs

Complexity in OM contexts has been dealt with more recently by framing operating systems as CASs (Choi et al. 2001; Surana et al. 2005; Pathak et al. 2007; Bozarth et al. 2009).

A CAS is a special class of complex systems that emerges over time into a coherent form, and adapts itself and emerges without any singular entity deliberately managing or controlling it (Holland 1995). Adaptation, self-organisation and co-evolution are the main features of CASs.

Adaption means that the system changes, improving its fitness for its environment and creates new forms of emergent order consisting in new structures, patterns and properties. Adaption is possible thanks to self-organisation, i.e. the new order arises from the interaction among agents without being externally imposed on the system (Goldstein 1999); self-organisation results in emergence, that is, a new order of some kind.

Self-organisation and emergence characterise the quasi-equilibrium state at the edge of chaos in which CASs operate, a state of non-complete order just short of chaos. It is a combination of regularities and randomness.

An important point emphasised by many authors is that CASs co-evolve with a changing environment. That is, the dynamic environment, by interacting with the CAS, forces changes in the entities that reside within it, which in turn induce changes in the environment (co-evolution). Kauffman (1993) observes that organisms do not merely evolve; they co-evolve both with other organisms and with a changing environment. He describes co-evolution as a process of coupled, deforming landscapes where the adaptive moves of each entity alters the landscapes of its neighbours.

Framing operating systems as CASs means recognising that the operating systems possess the characteristics of a CAS and that they behave as such. They are made up of a number of independent, multiple and heterogeneous human agents making decisions using personal heuristics and schemata. Interactions among the human agents allow the system to self-organise and co-evolve with the rugged and dynamic environment in which it exists.

CAS theory explains how CASs behave and how a new order emerges (Casti 1997; Johnson 2001). Thus, CAS theory applied to OM is aimed at explaining how heterogeneous agents in OM contexts "self-organize" to create new structures and at understanding how those structures emerge and develop.

3.2 CAS-Based Behavioural Operation Management Research

My rationale for doing BOM research is to build OM models applying methodologies to study CASs and to incorporate behavioural factors into these CAS-based models.

Following this logic, interesting research directions can be identified. Notice that research questions should be formulated coherently with those addressable by CAS theory.

Exemplar OM research questions addressable by using CAS theory are:

- How do operating systems evolve over time?
- How does variety in the elements affect evolution and the creation of order?
- How do strategies/decisions at single level impact the collective behaviour of the system?

- How does the topology of the pattern of interactions (i.e. due to the task interdependencies of a process, the supply chain structure, the product complexity, the technology, etc.) affect dynamics?
- How does the distribution of decision-making power affect performance?

Such questions can be formulated referring to anyone of the OM contexts mentioned above. Therefore, CAS theory allows one the main challenges of OM research to be faced: how to enrich and extend the body of knowledge on OM in different contexts by studying evolution and dynamism in operating systems, which is an issue currently lacking attention in OM literature (Pathak et al. 2007).

However, in order to do BOM research, behavioural factors including both individual and population properties should be incorporated into the research questions. Both descriptive and prescriptive research can be developed. As said above, the prescriptive approach analyses how operating system should work incorporating the behavioural factors. Instead, descriptive research is aimed at understanding the effect of behavioural factors on the decision-making process and in turn on the performances of operating systems (Gino and Pisano 2008).

For example, one could question:

- How do the operating system (e.g. process, firm and supply chain) evolves in the short and long run in the case of a particular decision maker's bias, such as overconfidence, conformism or anchoring heuristics?
- What new order emerges in the system in the case of individual cognitive biases?
- Is the emergence of a new order affected by a specific behavioural factor?
- Do the considered behavioural factors influence the resulting performances?

To give answers to the research questions above, CAS theory offers suitable tools and methods, as described in the next Section. Whatever the tool and method chosen, the advantage offered by CAS theory is that it permits the effect of behavioural factors and the resulting operating system behaviour to be studied as the spontaneous result of the system's self-organisation and co-evolution with the environment.

4 Complexity-Based Methodologies for BOM Research: *NK* Fitness Landscape

Recognising the complex nature of operating systems and studying them as CASs provide a rich set of tools to model and analyse their complexity. I limit attention to one methodology largely employed in complexity science literature for the study of CASs, i.e. *NK* fitness landscape.

It has been selected for a variety reasons. Firstly, it has been successfully applied in general management contexts to study organisational behaviour (Davis et al. 2007). It is well suited to behavioural research because it allows individual and social properties in OM models to be incorporated with ease, as I describe next. Furthermore, the proposed methodology is more suitable than experimental research, the main alternative methodology adopted in behavioural studies in many different fields such as economics, finance, organisation, for dealing with complexity. It is also better adaptive to overcoming the difficulty in predicting and understanding which individual agent strategies are most likely to lead to a desired collective behaviour. In OM contexts characterised by high complexity, experiments indeed could be costly and might not cover all possible events (Surana et al. 2005), while simulation may do this for a fraction of the cost.

Although extensively used in management studies, *NK* fitness landscape is novel in BOM. Moreover, there are even very few examples of applications in OM contexts in the literature (see for example Giannoccaro (2011) for the application of *NK* fitness landscape).

In the next section the methodology is described presenting key concepts and some common research questions which they are used to address. Finally, I discuss how the OM contexts and the behavioural factors might be modelled.

4.1 The NK Fitness Landscape

4.1.1 Description

The *NK* fitness landscape is a simulation technique advanced by Kauffman (1993) in the context of evolutionary biology and consists in a family of fitness landscapes which can be tuned by two parameters, *N* and *K*. In particular, the stochastic procedure proposed by Kauffman to design fitness landscapes has subsequently become popular in the modelling of organisational decision problems (Levinthal 1997; McKelvey 1999; Gavetti and Levinthal 2000; Rivkin 2001; Siggelkow 2001; Ethiraj et al. 2008; Ghemawat and Levinthal 2008; Ganco and Hoetker 2009; Giannoccaro 2011).

In these studies, the system (e.g. a firm, a product, a technology, a strategy, a plant, a supply chain), is conceptualised a set of *N* elements and *K* interactions. Each element may assume different states and, typically, it is assumed that each element occupies a binary state, i.e. 0 *or* 1.

The following describes applications of *NK* fitness landscape to the modelling of a firm. In such a case, the firm is conceptualised as a set of interdependent decisions. Decisions may concern what new product to be launched, the procurement policy, the production/transportation schedule, the inventory management policy, the adoption of IT, to name a just few possibilities.

A particular *N*-digit string $c = (d_1, d_2, ..., d_N)$ represents a specific combination of choices regarding the decisions to be made (configuration).

K is the average number of interactions among the decisions d_i. It models the richness of interactions among the decisions. Two decisions interact with each other when the contribution of a decision to the system payoff depends on the choice on the interacting decisions. For example, the adoption of a new IT is more or less effective on the basis of the decisions of the managers to invest in workforce formation; the decision to reduce the safety stock could be detrimental for the firm, if the decision on the procurement policy is to adopt a multiple sourcing policy under arm's length strategy conditions, whereas it could be much more effective in the case of a partnership and single sourcing policy.

The pattern of interactions among the decisions is contained in an $N \times N$ influence matrix where each x in the (i, j) position means that the column decision j influences the row decision i.

The different ways, in which the firm's choices (0–1) about the decisions are combined, generate 2^N possible configurations, to each of which is associated a fitness value for the overall system, i.e. a firm overall payoff $P(d)$. The map from each configuration into the overall payoff is the fitness landscape, where the position in the landscape corresponds to the configuration, and the height, to the payoff of the configuration.

The landscape is thus made up of valleys and peaks. The highest peak (global peak) corresponds to the configuration assuring the highest payoff. Local peaks are configurations with the highest payoffs in the neighboured positions (they correspond to the definition of local optima, while the global peak is the global optimum) and are good configurations.

A specific stochastic procedure is adopted to generate the fitness landscape. For each choice configuration, each single decision d_i offers a contribution C_i to the overall payoff, which in turn is calculated by averaging the N contributions. Therefore, $P(d) = [\sum_{i=1}^{N} C_i(d)]/N$. The contributions C_i are drawn randomly from a uniform distribution over $[0, 1]$. Note that each C_i depends not only on the corresponding decision but also on how the decisions interacting with it are resolved.

It is assumed that the firm is engaged in an adaptive walk across the landscape in search of the highest peak. The goal of the search is to identify the choice configuration that yields the highest firm overall payoff, or in other words, to reach the highest peak of the landscape (i.e. the global peak).

Thus, NK fitness landscape is an optimising decision making problem whose solution is achieved through the simulation of the firm's adaptive walk across the landscape.

The adaptive walk is simulated through a search algorithm. The most commonly adopted is based on an incremental improvement strategy consisting of the following steps: (1) new alternative configurations are proposed by changing a limited number of decisions, (2) all or some of the new alternative configurations are compared with the status quo, (3) a movement of the system to occupy the new configuration if better than the status quo.

A long jump strategy can be also employed (Levinthal 1997; Rivkin 2001), in which case, the system may jump directly (reproduce), with a specified degree of effectiveness (probability), to the best configuration.

The effectiveness of the search is a function of the shape of the landscape. The shape strongly depends on K. When $K = 0$, the landscape is smooth and single-peaked and the search for the global peak is very easy, even using an incremental improvement strategy. When K increases, the landscape becomes rugged and multi-peaked, i.e. with many local peaks, and the search for the global peak is less effective because the firm may be trapped in one of the local peaks.

The following steps should be followed to employ the NK fitness landscape methodology:

1. Fix N and K;
2. Fix the interaction matrix;
3. Generate the performance landscape;
4. Define the search algorithm;
5. Release the firm on the landscape;
6. Perform a search for the global peak;
7. Collect performances.

Performances in NK fitness landscape are measured in terms of efficacy and speed in finding the global peak. The efficacy of a search is commonly measured in terms of overall system payoff computed as a portion of the maximum performance attainable on the landscape. Performance both in long run (at the end of the simulation) and short run (first runs of the simulation) are collected.

Sometimes this performance is accompanied for explanatory purpose, or is estimated, by computing the number of local peaks and sticking points characterising a specific landscape. A local peak is a configuration such that no configuration differing by only one decision exists resulting in a higher payoff. The sticking point is a configuration of choices from which the system does not move, because there is no different configuration in one decision which meets the approval of the decision maker (Siggelkow and Rivkin 2002).

The higher the number of local peaks and the number of sticking points, the lower the efficacy of the system is likely to be since it has a high probability of becoming trapped into a sub-optimal configuration. The number of sticking point is also a measure of search diversity (or exploration), because the higher the number of sticking points, the higher the probability that the system will already be blocked into a sub-optimal configuration in the first stage of search, thus resulting in a low level of search diversity and exploration (Siggelkow and Rivkin 2005).

Search speed is measured as the average improvement in performance experienced during the first stages of a search.

Research questions that can be handled by employing NK fitness landscape should be framed as "problem solving" or, equivalently, a search for the optimal point on the landscape (Davis et al. 2007). Research questions are commonly formulated in the following terms:

- How long does the system take to find an optimal configuration?
- What is the performance of this configuration?
- What effect does increasing the number of elements (N) have on performance?

- What effect does increasing the level of interactions (K) have on performance?
- What effect does changing the overall pattern of interactions (interaction matrix) have on performance?
- How do performances change if alternative search algorithms are used?

In organisation studies the most relevant applications concern problems of organisational adaptation, organisational design and organisational behaviour. Specifically, in the latter area of study, they allow the consequences of cognitive limitations of the decision makers involved in the organisations to be analysed (Siggelkow 2011).

4.1.2 Coding OM Contexts into *NK* Fitness Landscape

Table 2 illustrates how different OM contexts might be modelled using *NK* fitness landscape. For each context what N, K, the pattern of interactions and the search algorithm model is described.

In the case of product development and R&D contexts, applications are suggested by explicitly referring to strategic and organisational studies (references are given in the last column of Table 2). I believe that since this context is cross-sectional to OM, strategic management, and organisational theory, applications in these fields could be considered as also pertaining to OM. In such a context, N and K are commonly employed to model the product and the technology in terms of the components and the number of interdependencies among them. The configuration of choices represents a specific product design or technology choice. The pattern of interactions describes the product architecture (e.g. modular versus integral). Incremental improvement algorithms are employed to model experiential learning, the wideness of search is used to model exploitation (limited search) and exploration (wide search), and long jump stands for reverse engineering practice or imitation of the leader.

For the other OM contexts I suggest possible ways of coding to identify which variables stand for N, K, the pattern of interactions and the search algorithm. To best of my knowledge, there are no applications of *NK* methodology in these OM contexts. The only exceptions refer to a study concerning the inventory and distribution system and a few applications to supply chain management.

4.1.3 Incorporating Behavioural Factors in *NK* Fitness Landscape

As discussed above, *NK* fitness landscape contributes to BOM through the development of CAS-based OM models incorporating behavioural factors. In the next section, I suggest how to code in *NK* fitness landscape the behavioural factors classified in (1) bounded rationality and cognitive biases, (2) social factors and (3) population factors.

Table 2 Coding OM contexts into *NK* fitness landscape

OM context	N	K	Pattern of interaction	Search algorithm	References
Product development and R&D	Number of components of product/technology	Degree of interdependence among components	Product architecture (modular, integral)	Long jump: Imitation by reverse engineering, exploration strategy; Incremental improvement: experiential learning, exploitation strategy	Rivkin and Siggelkow (2003, Siggelkow and Rivkin 2005) Gavetti and Levinthal (2000)
Project management	Number of activities/tasks making up project	Degree of interdependence among activities/tasks	Process pattern	Organisational structure of activities	
Inventory and distribution management	Number of stock keeping units, warehouse/DCs	Degree of interconnections	Distribution network	Level of centralisation	
Production management	Number of production sites, products, production tasks	Degree of process/product flexibility, i.e. shifting product production to another site, degree of task interdependence	Topology of the production sites, pattern of task interdependencies (pooled, sequential, reciprocal)	Production strategy	
Procurement and strategic sourcing	Number of suppliers, items to be purchased	Degree of cooperation among suppliers, degree of interrelations among items in bill of materials	Sourcing policy (multiple, single, and double sourcing)		

(continued)

Table 2 (continued)

OM context	N	K	Pattern of interaction	Search algorithm	References
Supply chain management	Number of supply chain firms, decisions	Degree of inter-firm relationships; density of links in supply chain	Type of integration problem; type of production process	Form of governance	Giannoccaro (2011) Choi and Krause (2006) Levinthal and Warglien (1999)

Bounded Rationality and Cognitive Biases

Bounded rationality of the decision maker means that people fail to be fully rational: they have limited cognition and computation capability to identify all the alternatives, determine all eventual consequences of each alternative and select the best according to the decision maker's preferences (Simon 1955).

Bounded rationality can be modelled in *NK* fitness landscape through the search capability of the agent. A local search algorithm, where only one decision is allowed to change, models high cognitive limitation of the decision maker. Instead, higher cognitive intelligence can be modelled by increasing the number of decisions that can be modified at the same time.

For example, to model search capability, Rivkin and Siggelkow (2003, 2005) introduce two variables: the search radius (SR) and the number of alternatives. The SR captures the degree of bounded rationality of the decision maker, namely the ability of the decision maker to make simultaneous changes to a wide range of decisions (Siggelkow and Rivkin 2006; Agarwal et al. 2011). A SR $= 1$ means that only one decision can be changed. The number of alternatives (ALT) a decision maker that may evaluate reflects the processing power of the decision maker at any organisational level (Siggelkow and Rivkin 2006). A large number of alternatives to be compared with the status quo, models a "smarter" decision maker.

Commonly, the decision maker selects randomly the alternatives to be evaluated. This selection, however, could be differently modelled for incorporating individual biases of decision maker. For example, the decision maker may generate alternatives at random and selected the ALT alternatives with the highest rank. The ranking algorithm is thus a modelling strategy for some individual cognitive biases. For example, confirmation bias could be modelled by a ranking algorithm giving higher rank to configurations having more similarities (i.e. number of decisions with the same choice) with the status quo.

More examples of how some individual biases could be modelled through the *NK* framework are given in Table 3. I suggest different modelling strategies, concerning not only search capability. They include:

- Search algorithm;
- Ranking algorithm for judging alternatives;
- Overall payoff calculation function;
- Decision rule to move, or not to move, into a new configuration;
- Generation of a new landscape along the same simulation.

I limit examples to the cognitive biases classified by Gino and Pisano (2008) and reported in Table 1. However, it should be noted that *NK* methodology allows different cognitive biases to be defined and modelled such as misperceptions of the decision maker about the degree of interactions, and pattern of interactions. In such cases, the modelling strategy adopted is the generation of a landscape different from the actual one, conforming to what the decision maker believes to be the degree and the pattern of interactions.

Table 3 Modelling individual properties through *NK* fitness landscape

Behavioural factor	Search algorithm	Ranking algorithm	Formula to calculate overall payoff	Decision rule for moving on the landscape	Landscape generation
Cognitive bound	Incremental improvement defined by search radius and number of alternatives				
Information avoidance			Average of C_i not including contributions referring to decisions i that are avoided		
Confirmation bias		Giving higher rank to alternatives more similar to status quo			
Availability heuristics		Giving higher rank to alternatives similar to certain past choices			
Salient information			Weighted average of C_i, with higher weight for certain decisions		
Procrastination				Jumping simulation step	
Anchoring and adjustment heuristics			Average of contributions referring only to most familiar decisions		
Representativeness heuristics					Modifying payoff in landscape so that similar configurations have similar payoffs

(continued)

Table 3 (continued)

Behavioural factor	Search algorithm	Ranking algorithm	Formula to calculate overall payoff	Decision rule for moving on the landscape	Landscape generation
Inconsistency			Using different formulas to calculate overall payoff during simulation		
Conservatism				To maintain status quo configuration even when better configuration is identified	
Overconfidence					Generation of landscape different from actual one
Wishful thinking				Movement to desired configuration, regardless of actual payoff	
Illusion of control	Developing alternatives external to search radius				
Misperception of feedback					No update of landscape when new events occurs

Social Factors

When a problem involves multiple decision makers, a complex web of interactions emerges among them, affecting how operating systems work and perform. The specific pattern of interactions are the result of the behavioural factors involved.

Thus, to incorporate the social factors in OM models it is necessary to model such a network of interactions, defining the procedure adopted by the multiple decision makers to coordinate among each other in the search for the global peak. This is defined by the organisational structure.

In particular, it is the firm's organisational structure that defines how decision makers coordinate their work to pursue a common goal. It defines the hierarchical position of the decision maker (middle manager or CEO), the responsibility in terms of what decision she/he controls and the level of centralisation of the decisions (Rivkin and Siggelkow 2003, 2005).

Thus, as a modelling strategy for social factors, I suggest inserting them in the organisational structure. They can either influence the formal organisational structure, or be responsible for the existence of an informal organisation structure. Both directions are feasible.

In an *NK* fitness landscape the organisational structure is modelled by defining: (1) how the configuration choices made by the different decision makers are reunified to propose alternative configurations of the overall system, (2) how they are ranked, and, (3) how they are selected. In fact, as there are multiple decision makers, the configuration of the overall system will be computed by merging the alternatives proposed by each decision maker.

For example, in case of a decision maker highly prone to collaborating with the others and strongly committed to the firm's success, rather than to evaluating and ranking the proposed alternatives on the basis of the economic return of the decision maker (local payoff), he/she may evaluate alternatives based on an overall firm payoff.

The fairness, trusting behaviour and reputation of a decision maker can be similarly modelled. For example, a decision maker may trust, or not, that a configuration choice proposed by another decision maker is correct and made in the best interest of the entire organisation, on the basis of the reputation of fairness of the latter. If he/she trusts the other decision maker, he/she could compute his/her own configuration choice using the configuration choice suggested by the other decision maker. Otherwise, if he/she does not trust the other decision maker, he/she may decide to ignore the other's decisions using a random choice configuration.

In the same way other social preference factors could be modelled, such as social status. The decision maker who is responsible for making the merging of local configuration choices suggested by the local decision makers and for proposing the overall system alternative choice (i.e. the CEO) which might prefer to rely more on the choices made by those decision makers with whom he/she has familiarity or those with a good social reputation, neglecting the others.

Alternatively, social factors may also be modelled as an informal organisational structure, which re-allocates decision-making authority to each decision maker on the basis of status, reputation, credibility, capabilities recognised by other individuals. This informal structure should be modelled parallel to the formal one. As a consequence, in the simulation decisions will be made by those decision makers informally empowered with decision-making authority in accordance with the informal organisational structure, regardless of the distribution of decision-making power in the formal one.

Properties of Groups and Populations

NK fitness landscape is a methodology well suited to describing and analysing population properties in operating systems, because it permits collective behaviours to be modelled as the spontaneous result of self-organised, multiple interactions among human decision makers. Population properties emerge over time without any control being exerted by the decision maker or by modeller.

Behavioural factors in an OM context which are particularly interesting to analyse include culture, knowledge and collective learning (Loch and Wu 2007). There are very few examples in the literature of *NK* models incorporating them.

I propose some modelling strategies to do this referring to the model put forward by Press (2008) who modelled the diffusion of best practice within groups of firms belonging to the same production segment of an industrial district. Each firm proposes its own alternative choice configuration and the payoff of each configuration is computed. The best one is selected and transferred to the other firms which then adopt it.

Culture acts as a norm to which each decision maker adheres without questioning it. Therefore, it can be modelled by assuming that the decision makers belonging to the same group or population adopt the same procedure to propose alternatives, the same ranking algorithm, the same formula to compute the overall payoff and/or the same rule regarding decisions to move to a new configuration.

Knowledge may be conceived in simulation as a repository of information accessible to any decision maker. In *NK* models, knowledge can be modelled through the performance landscape. The existence of a common base of knowledge means that decision makers use the same information to make decisions, i.e. the performance landscape. When a population is divided into groups, all the agents within a group have access to the same pieces of the entire body of information, therefore they only know some parts of the entire landscape.

4.1.4 BOM Research Through *NK* Fitness Landscape

What type of BOM research can be done by employing *NK* fitness landscape? Which BOM research directions can be suggested by resorting to *NK* fitness landscape?

Table 4 NK-based research questions for two OM contexts

NK-based research questions	NK-based OM research questions
New product development	
How long does it take to find an optimal configuration?	How much faster is the system at developing a new product/process?
What is the performance of this configuration?	How effective is the new product configuration?
What effect on performance does increasing the number of elements (N) have?	What effect on performance does increasing the complexity of the product in terms of number of components have?
What effect on performance does increasing the level of interactions (K) have?	What effect on performance does increasing the complexity of the product in terms of number of components have?
What effect on performance does changing the overall pattern of interactions (interaction matrix) have?	What effect on performance does making a product/process more/less modular have?
	Which product architecture offers a correct number of effective alternative configurations?
How do performances change using alternative search algorithms?	What effect on performance do different R&D strategies (e.g. exploitation versus exploration, first entry versus imitation) have?
Supply chain management	
How long does it take to find an optimal configuration?	How much faster is the system at finding an optimal supply chain configuration?
What is the performance of this configuration?	How effective is the supply chain configuration?
What effect on performance does increasing the number of elements (N) have?	What effect on performance does increasing the number of firms in the supply chain (i.e. reducing the level of vertical integration) have?
What effect on performance does increasing the level of interactions (K) have?	What effect on performance does increasing the number of links among firms have?
What effect on performance does changing the overall pattern of interactions (interaction matrix) have?	What effect on performance does pursuing modular or integral structures have?
	Which supply chain structure offers a correct number of effective alternative configurations?
How do performances change using alternative search algorithms?	What effect on performance do different forms of governance (e.g. centralisation versus decentralisation), have?

In order to answer these questions, we should formulate first OM research questions that are coherent with the use of NK fitness landscape. As discussed in Sect. 4.1.1, such research questions refer to how long the system takes to find optimal configurations and the performance of these optimal configurations. Moreover, the NK landscape fitness approach is suitable for analysing the effect on performance of the increase in the number of parts making up the system (N), the rise in the number of links among them (K), the change in the pattern of interactions ($[M]$) and the application of various search algorithms.

Table 5 Examples of BOM *NK* research questions in the supply chain management context

Prescriptive research questions

When decision makers are characterised by
high trust behaviour and reputation,

how much faster is the supply chain at finding an
optimal supply chain configuration?

how effective is the optimal configuration?

what effect on performance does increasing the
level of vertical integration have?

what effect on performance does increasing the
number of links among firms have?

what effect on performance do modular versus
integral structures have?

what effect on performance does the degree of
centralisation of the decision making power
have?

Descriptive research questions

Do trust behaviour and high reputation of
the decision makers influence:

the speed of the supply chain in finding an optimal
supply chain configuration?

the effectiveness of the optimal configuration?

to what extent increasing the level of vertical
integration affects performance?

to what extent increasing the number of links
among firms affects performance ?

to what extent the choice of modular versus integral
structures affects performance?

to what extent the degree of centralisation of the
decision making power affects performance?

Based on this, by adopting the coding of OM contexts shown in Table 3 in an *NK* fitness landscape, specific OM research questions addressable through *NK* fitness landscape can be formulated in any OM context.

Following this procedure, in Table 4 OM *NK*-based research questions are derived for two OM contexts, i.e. product development and R&D and supply chain management. The same questions might hold also for different OM contexts.

Notice that the suggested procedure is general. It may result in different research questions depending on the coding of the OM contexts in the *NK* fitness landscape.

Finally, in order to identify BOM research questions, it is necessary to incorporate behavioural factors into the *NK*-based OM research questions, coherently with both a prescriptive and a descriptive approach.

To do this, it is important to identify which are the most critical behavioural factors to study. I suggest making a list of behavioural factors for any OM context including both individual and population factors and evaluating the importance of each in terms of potential effect on performance in the specific context. If a behavioural factor does not affect how the operating system works and performs, it is not critical and it is not necessary to study it. In the same way, Gino and Pisano (2008) propose research questions by identifying for each of the considered contexts in which cognitive biases are more likely to occur and to affect performance.

Once the critical behavioural factors have been identified, BOM research questions can be formulated by questioning how behavioural factors affect performance and/or on how the system works and performs, given the included behavioural factors. Research questions should be formulated coherently with those addressable through *NK* fitness landscape.

For example, making explicitly reference to the supply chain management context, the behavioural factors affecting social interactions among decision makers are very relevant. The effective and efficient management of the supply chains indeed depends on how all decision makers interact among each other to reach a common goal. In such a context trust and reputation are key issues influencing the system dynamics (Ireland and Webb 2007; Johnston et al. 2004; McCarter and Northcraft 2007). Examples of BOM *NK* research questions are shown in Table 5.

The same approach can be adopted to generate BOM research questions addressable through *NK* fitness landscape including different behavioural factors by simply substituting "trust behaviour and reputation" in the first sentence of the research question with the factor to be considered.

5 Conclusions

There are few studies that have analysed methodological issues in BOM research. Although most BOM studies to date have used experiments, there is no need to constrain the study of how behavioural factors affect the work and the performance of operating systems only to this methodology.

In this chapter I have proposed the adoption of complexity science both from a theoretical and methodological point of view in BOM research. This argument has been discussed viewing operating systems as CASs, i.e. complex systems that are able to self-organise, adapt and co-evolve with a dynamic environment. Many dimensions of complexity are indeed recognisable in operating systems, such as the large number of parts making up the system, the variety of elements, the interconnections among them and the non linear and adaptive behaviours displayed by the system.

Moving from this assumption, I have suggested applying complexity science in order to study operating systems in diverse OM contexts and I have also proposed research questions coherent with a complexity science approach. They concern how operating systems behave, adapt and show new orders in terms of processes, structures and performances. This allows a gap in BOM literature to be filled. The latter has been focused more on adopting static approaches with scarce insights into the effect of behavioural factors on the dynamism and evolution of operating systems.

Is the emergent order affected by behavioural factors? Is the adaptivity of operating systems dependent on behavioural factors? When specific behavioural biases occur, which strategies are most effective in stimulating the system to evolve in a desired direction? Such research questions can be contextualised in any OM context and their answers will provide interesting new knowledge for the field.

I have also suggested the adoption of a simulation tool to study CASs to develop BOM models, i.e. *NK* fitness landscape. After reviewing the methodology and its main applications in organisational contexts, I have suggested how different OM contexts can be modelled and how behavioural factors both at an individual and at a population level might be operationalised through the methodology proposed.

Finally, I have suggested how to formulate research questions that might be addressed by applying *NK* fitness landscape. Using the proposed framework, interesting future research directions in BOM research in one OM context have been suggested.

In conclusion, this chapter wishes to offer two main contributions to BOM research: on the one hand, the suggestion that valuable BOM research is possible thanks to application of complexity science as a theoretical and methodological approach; on the other hand, the provision of a methodological toolkit based on *NK* fitness landscape allowing BOM models to be built.

References

V. Agarwal, N. Siggelkow, H. Singh, Governing collaborative activity: interdependence and the impact of coordination and exploration. Strateg. Manag. J. **32**, 705–730 (2011)

E. Bendoly, K. Donohue, K. Schultz, Behavior in operations management: assessing recent findings and revisiting old assumptions. J. Oper. Manag. **24**, 737–752 (2006)

C.C. Bozarth, D.P. Warsing, B.B. Flynn, E.J. Flynn, The impact of supply chain complexity on manufacturing plant performance. J. Oper. Manag. **27**, 78–93 (2009)

J.L. Casti, *Would-be Worlds: How Simulation is Changing the Frontiers of Science* (Wiley, New York, 1997)

T.Y. Choi, D.R. Krause, The supply base and its complexity: implications for transaction costs, risks, responsiveness, and innovation. J. Oper. Manag. **24**, 637–652 (2006)

T.Y. Choi, K. Dooley, M. Rungtusanatham, Supply networks and complex adaptive systems: control versus emergence. J. Oper. Manag. **19**, 351–366 (2001)

S. Chopra, W. Lovejoy, C. Yano, Five decades of operations management and the prospects ahead. Manag. Sci. **50**, 8–14 (2004)

J.P. Davis, K. Eisenhardt, C.B. Bingham, Developing theory through simulation methods. Acad. Manag. Rev. **32**, 480–499 (2007)

S.K. Ethiraj, D.A. Levinthal, R.R. Roy, The dual role of modularity: innovation and imitation. Manag. Sci. **54**, 939–955 (2008)

M. Ganco, G. Hoetker, NK Modeling Methodology in the Strategy Literature: Bounded Search on a Rugged Landscape, in *Research Methodology in Strategy and Management*, vol. 5, ed. by D.D. Bergh, D.J. Ketchen (Emerald Group Publishing, Bingley, 2009), pp 237–268

G. Gavetti, D. Levinthal, Looking forward and looking backward: cognitive and experiential search. Adm. Sci. Q. **45**, 113–137 (2000)

P. Ghemawat, D. Levinthal, Choice interactions and business strategy. Manage. Sci. **54**, 1638–1651 (2008)

I. Giannoccaro, Assessing the influence of the organization in supply chain management using NK simulation. Int. J. Prod. Econ. **131**, 263–272 (2011)

F. Gino, G. Pisano, Toward a theory of behavioural operations. Manuf. Serv. Oper. Manag. **10**, 676–691 (2008)

J. Goldstein, Emergence as a construct: history and issues. Emergence **1**, 49–72 (1999)

J.H. Holland, *Hidden Order* (Addison-Wesley, Reading, 1995)

R.D. Ireland, J.W. Webb, A multi-theoretic perspective on trust and power in strategic supply chains. J. Oper. Manag. **25**, 482–497 (2007)

S. Johnson, *Emergence: The Connected Lives of Ants, Brains, Cities, and Software* (Scribner, New York, 2001)

D.A. Johnston, D.M. McCutcheon, F.I. Stuart, H. Kerwood, Effects of supplier trust on performance of cooperative supplier relationships. J. Oper. Manag. **22**, 23–38 (2004)

S. Kauffman, *The Origins of Order: Self-Organization and Selection in Evolution* (Oxford University, New York, 1993)

D.A. Levinthal, M. Warglien, Landscape design: designing for local action in complex worlds. Organ. Sci. **10**, 342–357 (1999)

D.A. Levinthal, Adaptation on rugged landscapes. Manag Sci. **43**, 934–950 (1997)

C.H. Loch, Y. Wu, *Behavioral Operations Management* (Now Publishers Inc, Hanover, 2007)

M.W. McCarter, G.B. Northcraft, Happy together? Insights and implications of viewing managed supply chain as social dilemma. J. Oper. Manag. **25**, 498–511 (2007)

B. McKelvey, Avoiding complexity catastrophe in coevolutionary pockets: strategies for rugged landscapes. Organization Science,**10**, 294–321 (1999)

S.D. Pathak, J.M. Day, A. Nair, W.J. Sawaya, M.M. Kristal, Complexity and adaptivity in supply networks: building supply network theory using a complex adaptive systems perspective. Decis. Sci. **38**, 547–580 (2007)

K. Press, Divide to conquer? limits to the adaptability of disintegrated, flexible specialization clusters. J. Econ. Geogr. **8**(4), 565–580 (2008)

J.W. Rivkin, N. Siggelkow, Organizational sticking points on NK landscapes. Complexity **7**(5), 31–43 (2002)

J.W. Rivkin, Imitation of complex strategies. Manag. Sci. **46**, 824–844 (2000)

J.W. Rivkin, Reproducing knowledge: replication without imitation at moderate complexity. Organ. Sci. **12**, 274–293 (2001)

J.W. Rivkin, N. Siggelkow, Balancing search and stability: interdependencies among elements of organizational design. Manag. Sci. **49**, 290–311 (2003)

E. Siemsen, That thing called Be-Op's. POMS Chron. **16**, 12–13 (2009)

N. Siggelkow, J.W. Rivkin, When exploration backfires: unintended consequences of multi-level organizational search. Acad. Manag. J. **49**, 779–795 (2006)

N. Siggelkow, Firms as systems of interdependence choices. J. Manag. Stud. **48**(5), 1126–1140 (2011)

N.J. Siggelkow, J.W. Rivkin, Speed and search: design organizations for turbulence and complexity. Organ. Sci. **16**, 101–122 (2005)

H.A. Simon, A behavioral model of rational choice. Q. J. Econ. **LXIX**, 99–118 (1955)

H.A. Simon, The architecture of complexity. Proc. Am. Philos. Soc. **106**(6), 467–482 (1962)

A. Surana, S. Kumara, M. Greaves, U.N. Raghavan, Supply-chain networks: a complex adaptive systems perspective. Int. J. Prod. Res. **20**, 4235–4265 (2005)

Chapter 3
Trust in Face-to-Face and Electronic Negotiation in Buyer–Supplier Relationships: A Laboratory Study

Rossella Moramarco, Cynthia Kay Stevens and Pierpaolo Pontrandolfo

Abstract The purpose of this chapter is to study the role of pre-existing trust as a key factor for successful buyer–supplier relationships in electronic versus face-to-face negotiation mechanisms. It is known that e-sourcing can damage the buyer–supplier relationship, whereas face-to-face discussions can help elicit collaboration intentions and build trust. However, it is less recognized whether having established a prior trusting relationship can positively affect outcomes and strengthen the relationship even when electronic mechanisms are used. We explore such an issue by conducting a laboratory study which compares three negotiation mechanisms (i.e., face-to-face negotiation, e-mail negotiation, and e-reverse auction) across two pre-existing levels of buyer–supplier trust (i.e., high-trust and low-trust) in terms of their impact on perceived relational outcomes. Results confirm that higher pre-existing trust is linked to higher relational outcomes than low pre-existing trust; face-to-face negotiation is associated with higher supplier's perceived trust and satisfaction in dealing with the buyer compared to the e-mail negotiation and e-reverse auction. Furthermore, in the context of high pre-existing trust e-reverse auctions may not necessarily undermine existing relationships.

R. Moramarco (✉)
CEVA Logistics Italy, Strada 3, Assago, Milan 02 892301, Italy
e-mail: rossella.moramarco@gmail.com

C. K. Stevens
University of Maryland, Van Munching Hall, College Park,
Marryland 20742, USA
e-mail: cstevens@rhsmith.umd.edu

P. Pontrandolfo
Department of Mechanics Mathematics and Management,
Polytechnic University of Bari, Viale Japigia 182, Bari 70126, Italy
e-mail: pontrandolfo@poliba.it

I. Giannoccaro (ed.), *Behavioral Issues in Operations Management*,
DOI: 10.1007/978-1-4471-4878-4_3, © Springer-Verlag London 2013

1 Introduction

The adoption of different communication mechanisms for transacting business, such as face-to-face negotiation or e-sourcing via online negotiations and electronic reverse auctions, have been extensively researched. Empirical studies confirm that e-sourcing in particular can achieve cost reduction and procurement process improvements (Handfield and Straight 2003), although it has also been criticized for damaging the buyer–supplier relationship. The literature on interorganizational relationships provides a key to understanding why, as trust between the parties is an antecedent for collaboration, higher satisfaction, and reduced transaction costs. Face-to-face discussions can help to build trust whereas electronically mediated interactions provide a less strong basis for trust. Yet, it can be argued that a high pre-existing level of trust between buyer and supplier can positively affect outcomes and strengthen the relationship even when electronic mechanisms are used.

In this chapter we study the role of trust as a prerequisite for successful buyer–supplier relationships, and whether it mitigates the potential negative impact of e-sourcing on such relationships when compared to traditional face-to-face negotiations. The purpose of the paper is to investigate the independent as well as joint effects of trust and negotiation mechanisms on negotiation results. Specifically, we compare three negotiation mechanisms (i.e., face-to-face negotiation, e-mail negotiation, and e-reverse auction) across two pre-existing levels of buyer–supplier trust (i.e., high-trust and low-trust) in terms of their impact on relational outcomes. A set of hypotheses is developed using insights from the literature review and previous empirical suggestions.

We tested these hypotheses using a laboratory experiment in which MBA students played the role of buyers and suppliers seeking to reach agreement on a transportation service contract with multiple attributes (e.g., price, reliability, delivery interval). As shown by studies in many disciplines, including economics, psychology, and more recently purchasing and supply management, using an experimental design offers many advantages in identifying causal relationships. For example, randomly assigning participants to conditions ensures that individual differences such as personality or experience levels cannot explain the pattern of findings. Moreover, using the same background conditions and contractual specifications across all participants and sourcing mechanisms eliminates the influence of contract terms, market characteristics, and the competition dynamics as alternative explanations.

Transportation services represent an ideal context in which to study buyer–supplier transactions. Transportation services are required by businesses in many industries, although their features are not specific and may involve multiple important attributes (e.g., delivery time and reliability in addition to price). Because of this, the terms of each contract are likely to require negotiation and careful consideration, thereby reducing any benefits of prior experience when negotiating new deals.

The paper is organized as follows. In Sect. 2 we review the relevant literature related to negotiations and auctions mechanisms and trust in buyer–supplier relationships. Section 3 summarizes the main empirical findings regarding how sourcing mechanisms affect relationships. We then present the conceptual model and develop hypotheses in Sect. 4, and describe the experimental design and methodology in Sect. 5. Section 6 reports our findings and Sect. 7 discusses the study findings, limitations, and practical implications.

2 Literature Review

2.1 Negotiations and Auctions

Negotiation and auction mechanisms have gained considerable attention in recent years. Research in purchasing and supply management increasingly focuses on the benefits companies get from the adoption of such mechanisms.

A broad multidisciplinary definition has been provided by Bichler (2000): negotiation is an iterative communication and decision-making process between two or more agents (parties or their representatives) who:

1. Cannot achieve their objectives through unilateral actions;
2. Exchange information comprising offers, counteroffers and arguments;
3. Deal with interdependent tasks; and
4. Search for a consensus that is a compromise decision.

Bichler et al. (2003) highlight several additional features of a negotiation: the outcome, which can be an agreement or a disagreement; the negotiation arena, which is the place where negotiators communicate and interact; the agenda, which is the negotiation framework and includes specification of the issues to be discussed; the decision-making rules used to determine, analyze, and select alternative and concessions; and the communication rules, which determine the way offers and messages are exchanged. The negotiation protocol includes all the rules that define the negotiation arena, agenda, and permissible decision-making and communication activities of the negotiators. Note that, according to this definition, negotiations can occur either face-to-face or via mediating technology (e.g., phone, e-mail). In supply chain management, the focus tends to be on a single buyer and seller, although theoretically, a buyer could conduct negotiations sequentially with several different suppliers.

Negotiations can be contrasted with auctions, in which an individual or organization simultaneously considers offers to buy or to sell from multiple parties (i.e., individuals or organizations). Selling auctions are known as *forward* (i.e., the price tends to increase as the auction progresses), whereas purchasing auctions are called *reverse* (as the price tends to decrease during the action). In both cases, one party "controls" the market because supply and demand set the price and enable simultaneous comparisons across offers.

An electronic reverse auction (or simply reverse auction) has formally been defined as "an on-line, real-time dynamic transaction between a buying organization and a group of pre-qualified suppliers who compete against each other to win the business of supply goods or services that have clearly defined specifications for design, quantity, quality, delivery, and relayed terms and conditions" (Beall et al. 2003). They are characterized by short duration and constrained environments (there is not any possibility to provide detailed clarifications during the auction), thus having items clearly and unambiguously specified is of heightened importance.

Beall et al. (2003) argue that reverse auctions have been used for sourcing three of the four purchase categories of the Kraljic's matrix: noncritical, leverage, and bottleneck direct and indirect materials, including services and capital goods. Only for strategic purchases, which often involve long-term strategic relationships with suppliers and high switching costs, reverse auctions seem less appropriate and are rarely used (Handfield and Straight 2003).

Auctions and negotiations may influence the "soft" elements of a sourcing relation. For example, one of the direct consequences of using electronic auctions is a decreased commitment in the relationships by the supplier (Tassabehji et al. 2006). Suppliers perceive that auctions destroy their relationships with buyers (Jap 2003, 2007), in contrast to traditional negotiations which allow suppliers to develop *rapport* with the buying company—which means that mutual interest, positive feelings, and coordination emerge during the negotiation process (Huang et al. 2008). Because auctions and negotiations affect these "soft" elements, it may be important to consider the pre-existing trust between buyers and suppliers in order to understand how these sourcing mechanisms affect their subsequent relationships.

2.2 Trust in Buyer–Supplier Relationships

In the literature, various classifications and models of buyer–supplier relationships have been suggested. Losch and Lambert (2007) observe that there are several recurrent issues invoked to characterize such relationships and to identify intrinsic as well as extrinsic characteristics (Table 1). Intrinsic features (e.g., information exchange, trust, long-term orientation) describe how the parties characterize the relationship, whereas extrinsic characteristics represent the outcomes or results of the relationship (e.g., satisfaction and success). In general, extrinsic characteristics can be correlated with intrinsic qualities or the overall relationship characteristics.

Scholars agree that three broad types of buyer–supplier relationships (i.e., combinations of intrinsic and extrinsic characteristics) can be identified (AMR 1998): transactional, information sharing, and collaborative relationships. *Transactional relationships* involve short-term transactions set up for spot sourcing and entail operational activities carried out to execute the purchase, e.g., order request and receipt and payment. *Information-sharing relationships* involve frequent

Table 1 Characteristics of buyer–supplier relationships (*source* Losch and Lambert 2007)

Characteristics of buyer–supplier relationships	
Intrinsic characteristics	Extrinsic characteristics
Trust	Successfulness
Commitment	Satisfaction
Transaction-specific investments	Conflict level
Information sharing	
Long-term orientation	
Status of the relationship	
(new vs. incumbent)	

communications and data exchange about strategic as well as operational information regarding buyer's demand and supplier's offer, but they do not necessarily involve long-term collaborations to share specific knowledge and competencies. Finally, in *collaborative relationships*, buyers and suppliers jointly work to understand buyer's requirements, and information is used to develop customized solutions. The goal of collaborative relationships is to generate synergistic solutions to joint problems. Collaborative relationships often evolve into strategic partnerships or alliances, which involve mutual trust and commitment over an extended time period and a sharing of information as well as the risks and rewards of the relation (Simchi-Levi et al. 2000).

Among the various intrinsic characteristics, trust has been highlighted as the key feature of buyer–supplier relationships (Selviaridis and Spring 2007). Zaheer et al. (1998) give one of the most comprehensive definitions: *trust* is the expectation that an actor: (1) can be relied on to fulfill obligations, (2) will behave in a predictable manner, and (3) will act and negotiate fairly when the possibility for opportunism is present. Researchers posit that when interorganizational trust is high, agreements will be reached more quickly and easily; in presence of trust, parties are more flexible in granting concessions due to expectations that the other party will reciprocate in the future (Zaheer et al. 1998). Trusting relationships are also characterized by high level of information sharing: when a supplier can trust a given buyer, for example, the supplier will be more willing to share confidential information, such as production costs or product design and innovation (Panayides and Venus Lun 2009). Conversely, a lack of trust between the parties may prompt suppliers to withhold information that could be potentially useful for problem solving.

In addition, a positive relation between trust and transaction performance, defined as the outcome obtained at the end of the negotiation, has been found. Trust is a key success factor in improving innovativeness and supply chain performances (Morris and Carter 2004; Panayides and Venus Lun 2009), including costs reduction, delivery reliability, quality improvement, lead time, and flexibility. These findings suggest that firms may derive competitive advantage from relationships based on high levels of mutual trust.

3 Impact of Sourcing Mechanisms on Buyer–Supplier Relationships: Empirical Evidence

The impact of sourcing mechanisms and trust levels on interorganizational relationships is an emerging research issue. Most of the studies addressing this topic empirically analyze the way in which specific sourcing tools influence buyer–supplier relationships.

Gattiker et al. (2007) focus on suppliers' trust in buyers as an important outcome of e-sourcing adoption. They analyze how suppliers' trust varied under different procurement conditions, which depend on the type of sourcing mechanism and the complexity of the procurement situation. Their experimental study reveals that in reverse auctions, suppliers' trust levels are lower than in both face-to-face and e-mail negotiations.

Jap (2003, 2007) and Carter and Stevens (2007) study how the buyer's auction design (e.g., the number of bidders and the price visibility) affects the buyer–supplier relationship. Jap's studies provide empirical evidence that open-bid auctions result in greater supplier perceptions of buyer opportunism than do traditional sealed-bid formats. Suppliers generally dislike electronic reverse auctions because they feel that the computer interface prevents them from informing buyers about nonprice attributes, causing their products to become commoditized (Jap 2003). A laboratory experiment conducted by Carter and Stevens (2007) demonstrates that suppliers' perception of the buyer opportunism is increased when the auction shows suppliers' relative ranks (rank-based visibility) rather than the current lowest bid (price-based visibility); in addition, greater perceptions of opportunism are associated with including a larger number of suppliers in the auction (i.e., six versus three bidders).

Some data suggest that companies' experiences with electronic auctions may depend in part on how these interact with their own strategic orientations. For example, Caniels and van Raaij (2009) found that companies that compete on prices are very positive about electronic auctions; in particular, they worry less about the detrimental effects of such tools on their relationships. On the contrary, suppliers that seek to differentiate their offerings on the basis of innovative capabilities and excellent customer service report bad experiences with electronic reverse auctions and are less inclined to participate in future auctions. This is consistent with case studies and exploratory interviews conducted by Tassabehji et al. (2006) in food-packaging suppliers participating in reverse auctions. They report that many suppliers felt that existing relationships with buyers were significantly damaged following sourcing changes to reverse auctions: in particular, they resented being treated as commodity suppliers despite having contributed to buyers' product design and business development for many years. Suppliers reported suspecting that buyers sometimes entered "phantom" bids themselves to force price reductions and interview comments illustrated suppliers concerns—for example that "when a relationship is based on getting the lowest prices, mutual respect and value declines; there is now less trust and no feeling there is a partnership".

Taken together, these empirical findings suggest that suppliers' experiences of electronic sourcing adoption may cause them to lose trust with their customers in ways that undermine long-term collaborative relationships. Yet, little is known about how pre-existing trust levels influence suppliers' responses to ongoing use of electronic sourcing mechanisms and, in particular, how this compares with face-to-face negotiations.

4 Conceptual Model and Hypotheses

The literature review above discussed has emphasized that trust plays a key role in buyer–supplier relationships, fostering greater collaboration (Johnston et al. 2004) and richer information exchange. In the context of electronic negotiations there is potential for relationships to deteriorate, thereby diminishing performance (Handfield and Straight 2003); in such situations, trust becomes an important asset to buyers and suppliers. For example, Gattiker et al. (2007) argue that when using e-mail negotiation or e-reverse auction, suppliers' perceptions of buyer honesty positively affects suppliers' desire to have future interaction with the buyer. Similarly, Jap (2003) claims that low levels of trusts between the parties might lead to opportunistic behaviors, in turn causing lower performance in terms of product quality and service level.

We explore such issues by analyzing the impact of trust in the context of three different types of sourcing mechanisms: face-to-face negotiations, e-mail negotiations, and electronic reverse auctions. We anticipate that the intrinsic and extrinsic outcomes are influenced by two factors: (1) the type of mechanism used to transact business; and (2) the level of trust characterizing the buyer–supplier relationship. Most studies have considered the effects of sourcing mechanism and trust separately, perhaps implicitly assuming that richer media always facilitate trusting relations. Yet, even suppliers conducting face-to-face negotiation may not trust a given buyer due to "bad" prior experiences. An important question is whether the benefits derived from the transparency of face-to-face mechanism elicit improved collaborative intentions and propensity to continue the relationship. Similarly, it is unknown whether having established a prior trusting relationship can offset the reduced transparency of e-mail negotiations or electronic auctions.

Performances concern *relational* elements (Stank et al. 1999) and are related to "the extent of activities and behaviors directed toward initiating, developing, and/or maintaining successful industrial relational exchange" (Morgan and Hunt 1994). The following outcomes are measured in the study:

- *Satisfaction with dealing*: this results from evaluation of all aspects of the relationship between the parties (Sanzo et al. 2003; Benton and Maloni 2005; Ghijseen et al. 2009). It is a perception that follows the conclusion of a negotiation and influences future behavioral intentions, i.e., the likelihood that the parties will negotiate in the future (Oliver at al. 1994);

- *Expectation of continuity*: this measures the suppliers' intentions and expectations regarding a long-term relationship (Jap 2007). When a firm expects that the relationship will continue into the future, it is more willing to engage in processes and cooperate toward mutual beneficial solutions;
- *Desire for future dealing*: this represents the desire to transact business with the other party again in the future (Oliver et al. 1994) that contributes to measure the relational performance as well.

In addition, the development of the following relational factors is measured, that is the difference between the post-transaction and the pre-transaction level:

- *Trust*: trust has been previously defined as the expectation that the other party will behave in a predictable and reliable manner (Zaheer et al. 1998). As theoretical models and empirical findings show, trust acts as antecedent in enhancing collaborations and improving performance and is an important outcome of the relationship as well. Hence, pre-existing levels are taken into account when measuring performances, which enables measurement of changes due to the sourcing mechanism.
- *Perception of opportunism*: empirical studies reveal that suppliers often view auctions as being opportunistically employed by buyers. Given the definition of opportunism as "self-interest with guile" (Williamson 1985), the suppliers' perceptions of opportunistic behavior by the buyer are measured before and after the transaction, in addition to the broader concept of trust.

4.1 The Effect of Sourcing Mechanism on Relational Outcomes

According to media richness theory, one would expect to find that the information richness associated with different communication channels will affect the trust levels between the parties. In the initial relationship development, both buyers and suppliers regard face-to-face communication (for example meetings) with their counterparts as a necessary step to establish good working relationships (Ambrose et al. 2008). Richer communication media are better at transmitting complex and tacit knowledge and in supporting routine problem solving. Electronic media can be "relatively rich" if used in existing relationships (Vickery et al. 2004). Purchasing research on various electronic technologies (Ambrose et al. 2008) finds that buyers utilize information technology-based communication (e.g., EDI and e-mail) for tactical matters whereas they select richer modes (e.g., phone and face-to-face) for communication on less routine issues. Communication via e-mail is often used both when there is a great deal of uncertainty (in order to control the relationship), and when there is little uncertainty and the relationship has a low social content (in order to take advantage of convenience, ease of use, and speed).

In a negotiation context, negotiators are more likely to be collaborative when they use richer communication media. Participants who negotiate face-to-face (in experimental studies) report more trust in their opponents before and after the negotiation is completed compared to those using electronic communication (Naquin and Paulson 2003). Negotiators using richer communication media also express greater desire for future interactions than negotiators using leaner media. Less rich media make it easier for negotiators to mask the use of distributive bargaining tactics, thus possibly encouraging competitive behaviors (Purdy et al. 2000).

The experimental study conducted by Gattiker et al. (2007) reveals that the sourcing mechanism itself affects supplier's trust in buyer after the negotiation. First, face-to-face negotiation generally is associated with the highest level of supplier trust in buyer, followed by e-mail negotiation and e-reverse auction. "When trust-building is a critical outcome, there is no substitute for face-to-face" (Gattiker et al. 2007, p. 196). If buyers wish to use electronic tools when trust is an important outcome, they need to find alternative ways to establish trust, particularly in new relationships. Afterward, the authors match the sourcing tool with the complexity of the procurement and find that when procurement complexity is high, face-to-face negotiation and e-mail negotiation do not differ.

The use of electronic auctions has been linked to a distributive form of negotiation (Kaufmann and Carter 2004) since it tends to result in "pie expansion" (Jap 2003). Consistently with transaction economics insights, Beall et al. (2003) find that many suppliers perceive that electronic auctions negatively affected their relationships with their customers and lead to lower level of trust. The latter perspective is confirmed by Jap (2003), who argues that auctions are ideally suited for transactional exchange contexts but may be less appropriate for relational exchanges. Auctions—in particular those based on price competition—may inhibit collaboration in relational contexts (Emiliani and Stec 2004) because such auctions do not allow the expression of non-price attributes, such as quality, service, and reliability. The buyer's choice to use a face-to-face or electronic-mediated mechanism may encourage the supplier's suspicion that the buyer is using the auction opportunistically against the supplier.

We sought to replicate Gattiker's findings concerning impact of sourcing mechanism richness on the relational development—measured by the potential increase of trust (and decrease of perception of opportunism) from pretransaction phase to post-transaction phase.

Hypothesis 1. Increasing "richness" of the sourcing mechanism (i.e., electronic reverse auction, e-mail negotiation, or face-to-face negotiation) will positively influence the relational outcomes (i.e., satisfaction with dealing, expectation of continuity, and desire for future dealing) as perceived by the supplier. In particular:

Hypothesis 1a. Face-to-face negotiation will result in higher relational outcomes for suppliers than will e-mail negotiation.

Hypothesis 1b. Face-to-face negotiation will result in higher relational outcomes for suppliers than will e-reverse auction.

Hypothesis 1c. E-mail negotiation will result in higher relational outcomes for suppliers than will e-reverse auction.

4.2 The Effect of Trust on Relational Outcomes

The ultimate outcomes of a given buyer–supplier transaction can also be affected by the pre-existing trust between the parties. Specifically, pre-existing trust levels reduce uncertainty about the other party's motives, lessening the likelihood of miscommunication and misunderstandings that may damage the relationship. Benton and Maloni (2005) empirically find that supplier satisfaction is not affected by the final performance of the buyer–supplier transaction but instead results from the nature of the ongoing relationship as measured by trust, cooperation, and commitment. Experimental studies confirm that trust positively influences satisfaction with the relationship (Andaleeb, 1996). An increased level of future dealing expectation has been revealed from negotiators who experienced trust and collaborative interactions (Purdy et al. 2000; Naquin and Paulson 2003). In contrast, lack of prior trust may have a negative impact on propensity to continue the relationship and desire for future dealings.

Therefore, the following hypothesis is defined:

Hypothesis 2. The level of pre-existing trust between the buyer and the supplier has a positive impact on the relational outcomes (i.e., satisfaction with dealing, expectation of continuity, and desire for future dealings). That is, a high level of pre-existing trust will result in higher satisfaction with dealing, higher expectation of continuity, and higher desire for future dealing than will low level of pre-existing trust.

4.3 Potential Interaction of Sourcing Mechanism and Trust on Relational Outcomes

The use of electronic mechanisms in situations characterized by low levels of pre-existing trust may cause further deterioration of relational outcomes relative to the use of face-to-face mechanism. That is, the negative effects of lean communication media linked to sourcing mechanisms might be especially damaging in the presence of pre-existing mistrust, whereas it might be offset by a high level of pre-existing trust. Conversely, an initial absence of pre-existing trust might be counterbalanced by the trust-enhancing effects of rich media in the face-to-face negotiation so that relational outcomes are repaired.

In e-reverse auctions, relational outcomes are expected to be lower than they are in face-to-face and e-mail negotiations at any level of pre-existing trust, since the characteristics of the mechanism (lack of interpersonal contact) provide few avenues for interaction and cannot be mediated by the pre-existing level of trust. Less rich media make it easier for negotiators to mask the use of distributive bargaining tactics, thus possibly encouraging competitive behaviors (Purdy et al. 2000). E-mail negotiation has been associated with lower level of pre-existing trust (and post-contact trust as well) than face-to-face negotiation (Naquin and Paulson 2003).

Hence, we predict a significant interaction between the pre-existing level of trust and the richness of the sourcing mechanism. We explore the effects of each mechanism separately for both high and low trust.

Hypothesis 3a. Face-to-face negotiation will result in higher relational outcomes among suppliers (satisfaction with dealing, expectation of continuity, and desire for future dealing) than will e-mail negotiation when the pre-existing level of trust is low; when the pre-existing level of trust is high face-to-face will result in lower relational outcomes than will e-mail negotiation.

Hypothesis 3b. Face-to-face negotiation will result in higher relational outcomes among suppliers (satisfaction with dealing, expectation of continuity, and desire for future dealing) than will e-reverse auction when the pre-existing level of trust is low; when the pre-existing level of trust is high face-to-face will result in lower relational outcomes than will e-mail negotiation.

Hypothesis 3c. E-mail negotiation will result in higher relational outcomes (satisfaction with dealing, expectation of continuity, and desire for future dealing) among suppliers than will e-reverse auction at both high and low levels of pre-existing trust.

We analyze the interaction effect across pre-existing trust levels for each sourcing mechanism separately. However, we posit that the level of pre-existing trust will have a positive influence on the relational outcomes for all three sourcing mechanisms:

Hypothesis 4a. The level of pre-existing trust has positive impact on the relational outcomes (satisfaction with dealing, expectation of continuity, and desire for future dealing) among suppliers when face-to-face negotiation is used as a sourcing mechanism. That is, the relational outcomes in high-trust condition are higher than the relational outcomes in low-trust condition when face-to-face negotiation is used.

Hypothesis 4b. The level of pre-existing trust has a positive impact on the relational outcomes (satisfaction with dealing, expectation of continuity, and desire for future dealing) among suppliers when e-mail negotiation is used. That is, the relational outcomes in high-trust condition are higher than the relational outcomes in low-trust condition when e-mail is used.

Hypothesis 4c. The level of pre-existing trust has a positive impact on the relational outcomes (satisfaction with dealing, expectation of continuity, and desire for future dealing) among suppliers when e-reverse auction is used. That is, the relational outcomes in high-trust condition are higher than the relational outcomes in low-trust condition when e-reverse auction is used.

4.4 Trust Development

A further interesting aspect is related to *trust development* during finalization of contract terms, namely the potential increase of trust and decrease in *perceptions of opportunism* from the pre-transaction to the post-transaction phase. With regard to sourcing mechanisms, previous experimental findings (Huang et al. 2008) reveal that suppliers' trust in buyers (i.e., their perceived honesty and benevolence) grows significantly during face-to-face negotiations, whereas when e-mail is used an increase of honesty is observed only for complex procurement situations (i.e., those for which multiple and ambiguous attributes of the exchanged good or service are discussed in the procurement process). In electronic reverse auction the level of trust does not change during the sourcing event.

Based on these findings, we propose the following set of hypotheses:

Hypothesis 5. The "richness" of the sourcing mechanism has a positive impact on the development of suppliers' perceived trust in buyers following completion of the deal. In particular:

Hypothesis 5a. Face-to-face negotiation will result in increased perceived trust and decreased perceived opportunism from the pre-transaction to post-transaction phase.

Hypothesis 5b. E-mail negotiation will result in increased perceived trust and decreased perceived opportunism from the pre-transaction to post-transaction phase.

Hypothesis 5c. Electronic reverse auction will result in decreased perceived trust and perceived opportunism from the pre-transaction to post-transaction phase.

In addition, pre-existing trust may serve as foundation to foster the development of additional trust. Compared with low levels of pre-existing trust, high levels of pre-existing trust make it more likely trust will increase and suspicion of opportunistic behavior will decrease during the transaction. Alternatively, it has been argued that "high levels of existing trust may decrease the amount of future trust that can be formed because high levels of existing trust leave less room for new trust to grow (conceptually, as well as merely methodologically)" (Huang et al. 2008, p. 69).In other words, high pre-existing trust may create a "ceiling effect" that limits any further increases in trust. Given this divergence in reasoning, we considered two competing hypotheses regarding the effects of pre-existing trust on trust development:

Hypothesis 6. The level of pre-existing trust between buyer and supplier will affect the development of relational outcomes.

Hypothesis 6a. The level of pre-existing trust between the buyer and the supplier has a positive impact on the development of the relational outcomes. In particular, perceived trust will grow when the pre-existing level of trust is high and decrease when the pre-existing level of trust is low, and perceived opportunism will decrease when the pre-existing level of trust is high and grow when the pre-existing level of trust is low.

Hypothesis 6b. The level of pre-existing trust between the buyer and the supplier has a negative impact on the development of the relational outcomes. In particular, perceived trust will decrease when the pre-existing level of trust is high and grow when the pre-existing level of trust is low and perceived opportunism will grow when the pre-existing level of trust is high and decrease when the pre-existing level of trust is low.

The potential interaction between sourcing mechanism and trust condition and their effects on changes in perceived trust opportunism will be analyzed as post hoc research questions.

5 Methodology and Experimental Design

5.1 Behavioral Laboratory Experiments

The purpose of the traditional research on purchasing and supply management is the development of models and techniques that help decision makers to optimize their decisions. However, such models rely on specific assumptions that may not hold when implemented in real settings (Bendoloy et al. 2006), for example unbounded rationality and risk neutrality. Recent studies addressing this topic have proposed that new approaches are necessary to overcome the limitations of the traditional assumptions. *Behavioral operations management* (Gino and Pisano 2008) has the purpose of taking cognitive limitations, perceptions and personal attributes of individuals into consideration. Consistently, *behavioral experiments* are a suitable tool for the empirically study of behavioral operations issues (Bendoly et al. 2006; 2008). They are aimed at investigating relationships by manipulating treatments to determine the exact effect of controlled and independent variables on specific dependent variables (Wacker, 1998).

We undertook a laboratory (lab) experiment to test our hypothesized relationships. For the purposes of this research, a *laboratory experiment* is defined as a study involving participants that occurs in an environment that has been created for research objectives as a stylized version of a real setting (Colquitt 2008). The assumption underlying laboratory studies is that theory being tested applies in real-world situations and to actors outside the laboratory. Their primary advantage is a high degree of control over threats to internal validity, namely extraneous confounding factors that might affect the inference of causal relationships between the independent and dependent variables (Campbell and Stanley 1963). For example, in a field study, numerous differences among auctions (such as number of bidders, level of information visibility, and duration) could influence participants' confidence in the sourcing mechanism and thus mask the relationship between sourcing mechanism, trust and performances. By contrast, we standardized the contract characteristics and the cost structures among suppliers and buyers, thereby holding constant the purchase and market characteristics. Such tight control yields a high degree of confidence that observed differences in the outcome variables are actually due to differences in the independent variables.

One disadvantage of laboratory experiments is the lack of contextual realism, since they use students instead of real representatives of the population under study. However, as noted by Bendoly et al. (2006), "this can be a valid criticism if the phenomena under study heavily depend on the individual life experiences of the subjects". As noted earlier, this consideration was not relevant here, since even experienced suppliers must attend closely to the terms of unique transportation contracts. Laboratory experiments thus provide a valuable complement to the existing field studies by providing highly controlled environments to test the causal effects of independent variables. Although some facets of external validity may be compromised, the tradeoff in increased control affords researchers a sound basis for inferring causal relationships.

5.2 Experimental Design and Setting

This research explores how different conditions in face-to-face and electronic-mediated sourcing mechanisms affect transaction performance in logistics services procurement. In particular, we compared three sourcing mechanisms (face-to-face negotiation versus e-mail negotiation versus electronic reverse auction) at two levels of pre-existing trust between the buyer and the supplier (*high trust* versus *low trust*), resulting in a 3 × 2 experimental design in which participants are randomly assigned to one of the six resulting conditions(or treatments). The experimental design is depicted in Fig. 1.

The dependent variables are the relational performances described in Sect. 4. We measure the suppliers' perception of the relationship at the end of the transaction (i.e., satisfaction with dealing, expectation of continuity, and desire for future dealing) as well as pre-to post transaction changes in the relationship (i.e., changes in perceived trust and opportunism).

Our sample consisted of 95 MBA (93 %) and business Ph.D. students (7 %) enrolled in graduate courses at a large mid-Atlantic university. Of these, 63 % were male, 73 were U.S. citizens, and the ethnic breakdown was as follows: 55 % Caucasian, 32 % Asian, 6 % African American, 4 % Latino, and 3 % other. Subjects reported an average age of 28.38 years (S.D. = 4.16; range 24–36). Participants took part in the study in exchange for either extra course credit or gift certificates.

It is worth noting that although the sample included graduate students rather than experienced suppliers, all respondents had prior professional work experience, which reduces concerns about sample representativeness. Some existing

Fig. 1 Experimental design

Independent Variables	
Trust	**Sourcing mechanism**
High	Face-to-face negotiation
	E-mail negotiation
	E-reverse auction
Low	Face-to-face negotiation
	E-mail negotiation
	E-reverse auction

Dependent variables
Satisfaction with dealing. Expectation of continuity. Desire for future dealing. Trust. Perception of opportunism.

research suggests that MBA students and executives show highly similar patterns in organizing information and making decisions (Croson and Donohue 2006). Thus, the use of graduate business students in this study should not lead to dramatically different results than would be observed with experienced suppliers. Moreover, as business majors, students in the sample are representative of the larger population from which buyers and sales representatives are typically drawn.

In order to test the experimental design depicted in Fig. 1, the two independent variables (sourcing mechanism and trust) were manipulated and six different treatment conditions were defined. For *sourcing mechanism,* the manipulation consisted of assigning participants to one of the three conditions in interacting with buyers: (1) face-to negotiation, (2) e-mail negotiation, and (3) electronic reverse auction. For *trust* between the buyer and supplier, two levels were manipulated: high trust and low trust, which was operationalized by creating two versions of the scenario indicating pre-existing trust between the buyer and the supplier. The trust manipulation was generated based on insights from the literature review on the construct definition and pilot tested in an online study involving separate sample. For example, it is known that trust develops between partners over time and is closely tied to past experiences (Young-Ybarra and Wiersema 1999; Tian et al. 2008). Therefore, this firm-specific information concerning prior relationships provides evidence about the trustworthiness of the exchange partner.

Within a given sourcing mechanism condition, all participants pairs were assigned to either the high or the low trust condition (i.e., there were no circumstances in which one party read a scenario indicating high pre-existing trust and was then assigned to interact with a party who had read a scenario indicating low pre-existing trust). Moreover, to ensure that participants' perception of trust level was consistent with the manipulation, several manipulation checks were included in the preliminary survey.

For the most part, the buyer and seller scenarios contained identical information, however, as with many simulated negotiations, buyers were provided with unique facts about their own preferences that were not provided to sellers, and vice versa.

5.3 Experimental Procedures

Standardized written procedures were developed for all treatments-related aspects of the study, such as creation of material packets, communication with subjects, participant recruitment, and organization of sessions (Cook and Campbell 1979). Potential participants were recruited in their graduate classes and offered information about the scope of the study, benefits and risks of participation, and incentives (i.e., extra course credit, gift certificates). Using the list of students who signed-up to participate on particular day and time, participants were randomly assigned to the six treatments previously described.

Subjects were randomly assigned to one sourcing mechanism once they arrived at the laboratory. In the face-to-face and e-mail negotiation conditions, subjects were randomly assigned to be either the buyer or the supplier and then randomly paired with each other. Each dyad was randomly assigned to either the high or low trust condition. For the reverse auction sessions, suppliers were told that they would bid against other competing suppliers; however, they actually used an auction simulator and bid against the computer. Using a simulator eliminated the possibility of confounding factors, e.g., differences in competition dynamics from auction to auction. The simulator has the look and feel of a real online auction, and the computerized logic mimics the behavior of competing bidders. The simulator was programmed using Microsoft Visual Basic and developed on the price-based auction simulator used by Gattiker et al. (2007). Students who participated in the auction as suppliers were also randomly assigned to either the high or low trust condition.

After all participants arrived, they watched standardized powerpoint presentation that provided an overview of the study, described the procedure, and allowed time for them to review their role materials and complete the preliminary survey. The preliminary survey contained manipulation check items to test participants' perception of trust manipulation.

Dyads in the face-to-face condition were instructed to meet for a maximum of 60 min to conduct their negotiation. E-mail dyads were provided with appositely created e-mail addresses (i.e., participants' real names were not used) and instructed to negotiate using e-mail over the next 60 min, with the first e-mail to be sent within 5 min. All suppliers in face-to-face and e-mail conditions were instructed to make the first offer to the buyer. Suppliers in reverse auctions received written instructions and practice training in the use of the auction system. Then, they were instructed to log in the application and start bidding. An initial time length of 5 min was set for the auction, consistent with prior research showing that longer auction periods yield no differences in bids or bidding patterns (i.e., most bidding occurred during the first and last two minutes of the auctions, regardless of how long the actual auction period was; Carter and Stevens 2007). However, a soft-closing time was designed, namely the auctions extended for 90 s every time a bid was placed during the last 45 s; the purpose of extending the auction in this way was to possibly avoid sniping behaviors by bidders (i.e., placing bids in the last few seconds of the auction) (Chen-Ritzo et al. 2005).

At the conclusion of the transaction, participants were asked to complete a follow-up survey that recorded the settlements terms and individual perceptions on relational performance. Each experimental session concluded with a debriefing.

5.4 Contract Scenario

Participants received packets of written information that described their roles in the experiment. Two roles were assigned, depending on the sourcing mechanism condition: a buyer for the company desiring to purchase transportation services or a sales representative for the supplier company (note: only the supplier role was provided to participants assigned to the electronic auction condition). The same hypothetical transportation service contract was designed for use in all experimental conditions. Due to the complex nature of the logistics service, a multi-attribute contract was used to provide a clear specification of contract terms. This choice was consistent with the critical issues in the procurement of services (Andersson and Norrman 2002).The buyer was described as an industrial manufacturer that needed to purchase transportation services for the movement of final products (computer and technical equipment) from production facilities to central warehouses and then to distributors and retailers. The supplier was depicted as a logistics services provider that offered global, value-added, customized logistics solutions.

After a brief introduction containing the names of the organizations and their representatives, the background information contained the trust manipulation, which described information about the prior relationship between the two parties. In the *high trust* scenario the relationship between the two companies was characterized by honesty, reliability, positive past experiences, and further features that induced the parties to conclude that the opposite party could be trusted. In the *low trust* scenario, all the inductions were reversed in a way that the relationship was characterized by dishonesty, unreliability, and bad past experiences.

Then, the role material provided a detailed description of the contract terms, confidential information about the company's preferences concerning the service attributes, and guidelines regarding how to conduct the transaction. The contract had three service-related attributes of interest: price p, delivery interval d, and reliability r. *Delivery interval* is the time interval between two deliveries; it is expressed in time units (e.g., a delivery interval of six days). Higher values of delivery interval mean lower frequency of deliveries. *Reliability* is the expected rate of on-time deliveries (e.g., 95 % of scheduled transports expected to be provided on time). *Price* is related to the total price of the contract that the buyer pays to the provider for purchasing the transportation service. To aid their decisions concerning offers and counteroffers, suppliers and buyers were given a table showing the costs and the utilities, respectively, associated with each level of delivery interval and reliability.

5.5 Measures

In order to measure the relational aspects concerning the relationships between the buyer and the supplier, questionnaires were administered in pre-test and post-test phases of the experiment. As suggested by Johnston et al. (2004), studies on inter-firm relationships typically use individuals' reports to assess the perception of the relationship at the interorganizational level. Although a few studies have looked at both interpersonal and interorganizational trust in buyer–supplier relationships (Zaheer et al. 1998), in this study the more conventional approach of using participants' assessments to represent their organizations perceptions and relational attitudes toward the other party is adopted.

Table 2 Reliability
Cronbach's alpha for
multi-item scales

Variable	Cronbach's alpha
1. Pre-test trust	0.974
2. Pre-test opportunism	0.961
3. Post-test trust	0.936
4. Post-test opportunism	0.883
5. Satisfaction with dealing	0.914
6. Expectation of continuity	0.953

Multiitem scales measured on a 7-point Likert scale (1 = "strongly disagree" and 7 = "strongly agree") were adapted from existing research: "strongly disagree" denoted poor performance and "strongly agree" strong performance.

Trust was measured using the scale developed and validated by Doney and Cannon (1997) that measures interorganizational trust between buyers and suppliers firms. The measure of *perception of opportunism* was derived from the experimental study of Carter and Stevens (2007), who adapted existing items (e.g., Morgan and Hunt 1994) to the context of their study based on the field research of Beall et al. (2003) and Jap (2003). *Satisfaction with dealing* was measured by using the items from Ghijsen et al. (2009); the items measuring *expectation of continuity* were derived from Jap (2007), whose work assesses suppliers' confidence in future relationships in the context of electronic auction usage. The scale for *desire for future dealings* included a single item used by Gattiker et al. (2007), who adapted the measure developed by Oliver (1994).The items used for the measures are listed in appendix.

For each variable the items were averaged to form a composite measure. To assess the degree to which the items are free from random error and measure the construct in a consistent manner, reliability analysis is suitable. Reliability is typically assessed using Cronbach's alpha coefficient. A scale is found to be reliable if the coefficient is 0.70 or higher. In this study, reliability analysis was performed with the collected data for the pre-test and post-test multiitems measures. An excellent level of inter-item agreement was achieved for all the measures, namely a Cronbach's alpha higher than 0.85 was achieved for all the scales (Table 2).

6 Analysis of Results

6.1 Sample

The study involved a 3 (face-to-face negotiation versus e-mail negotiation versus e-reverse auction) × 2 (high trust versus low trust) experimental design. Although data were collected from both buyers and suppliers, our analyses here reported focus on the suppliers' responses (recall that no buyer data could be collected in the e-reverse auction condition). Therefore, the considered sample size is 65 suppliers' responses. Of those, four responses were omitted from the analysis since

Between-subjects Independent Variables		Cell size (n)	Within-subjects Dependent Variable	
Trust (high versus low)	Mechanism (face-to-face versus e-mail versus e-reverse auction)		Pre-test phase	Post test phase
High	Face-to-face negotiation	8	DVs: Trust and Perceived Opportunism	DVs: Trust and Perceived Opportunism
	E-mail negotiation	9	DVs: Trust and Perceived Opportunism	DVs: Trust and Perceived Opportunism
	E-reverse auction	13	DVs: Trust and Perceived Opportunism	DVs: Trust and Perceived Opportunism
Low	Face-to-face negotiation	9	DVs: Trust and Perceived Opportunism	DVs: Trust and Perceived Opportunism
	E-mail negotiation	10	DVs: Trust and Perceived Opportunism	DVs: Trust and Perceived Opportunism
	Electronic reverse auction	12	DVs: Trust and Perceived Opportunism	DVs: Trust and Perceived Opportunism

Fig. 2 Cells size and mixed-design repeated measures MANOVA

they reported results that were beyond the reasonable range of final profit given the scenario provided in the background information. A problem in research using human subjects is that some participant may neglect to do their tasks conscientiously. This appears to have been the case with these few responses. Omitting the unusable data points leaves 61 observations for the hypothesis tests. Sample sizes in each cell vary from 8 to 12 (Fig. 2), which are adequate for a laboratory experiment design (Hair et al. 1998) as suggested by recent experimental research (Gattiker et al. 2007; Carter and Stevens 2007).

6.2 Models of Analysis

Trust and perception of opportunism were measured in the pre-test as well as in the post-test phases, therefore a mixed-design approach was adopted for testing the hypotheses concerning those variables. A mixed-design analysis of variance separately examines the effect of between-subjects factors (i.e., independent variables in which a different group of subjects is exposed to each treatment condition) and within-subjects factors (i.e., often referred to as "repeated-measures variables" since more than one measurement is taken from each subject). In this study, sourcing mechanism and trust are the between-subject variables and the *test time* (pre-test and post-test) is the within-subject factor. The repeated-measures MANOVA design is depicted in Fig. 2. Means, standard deviations, and intercorrelations for study variables are shown in Table 3.

Table 3 Means, standard deviations, and correlation of dependent variables

Variable	Mean	SD	1	2	3	4	5	6	7
1. Pre-test trust	3.80	2.13	1						
2. Pre-test opportunism	4.06	1.75	−0.913**	1					
3. Post-test trust	4.39	1.30	0.691**	−0.734**	1				
4. Post-test opportunism	3.40	1.48	−0.592**	0.574*	−0.820**	1			
5. Satisfaction with dealing	5.40	1.25	0.527**	−0.564**	0.808*	−0.658**	1		
6. Expectation of continuity	5.17	1.51	0.553**	−0.567**	0.828*	−0.691**	0.861**	1	
7. Desire for future dealing	5.55	1.24	0.420**	−0.506**	0.814**	−0.697**	0.806**	0.862**	1

*$p < 0.05$, two-tailed
**$p < 0.01$, two-tailed

6.3 Validation of Measures

To ensure that participants perceived the trust manipulations as intended, a t-test on items taken prior to the start of the contract negotiations or auctions showed significant differences in the predicted direction [$t(60) = 13.99$, $p < 0.0001$; low trust $M = 1.79$, high trust M $= 5.88$]. These results indicate that the trust manipulation was successful, namely the level of perceived pre-existing trust between the buyer and seller was significantly higher in the *high trust* condition than in the *low trust* condition.

6.4 Results for Relational Outcomes

Hypothesis 1 predicted that the richness of the sourcing mechanism would positively influence relational performance such that the three mechanisms would follow a pattern in which face-to-face negotiations would result in higher relational outcomes than would e-mail (H1a) or e-reverse auctions (H1b) and e-mail negotiations would result in better outcomes than e-reverse auctions (H1c).

Consistent with Hypothesis 1, an ANOVA test (Table 4) on satisfaction with dealing indicates a significant main effect of sourcing mechanism [$F(2, 55) = 15.538$, $p < 0.001$; $\eta^2 = 0.361$; power $= 0.999$]. The pairwise comparisons across the three mechanisms (Table 5) show the predicted pattern with regard to satisfaction with dealing: satisfaction in electronic reverse auctions ($M = 4.705$) is 1.026 points lower than in e-mail negotiation ($M = 5.731$) and 1.357 points lower than in face-to-face negotiation ($M = 6.063$); however, means in the face-to-face and e-mail negotiation do not significantly differ.

Similar results derive from the MANOVA on expectation of continuity and desire for future dealings. The multivariate tests (Table 6) indicate a significant main effect of sourcing mechanism on these dependent variables [Pillai's trace $= 0.251$, $p < 0.01$, $\eta^2 = 0.125$, power $= 0.887$]. The univariate tests (Table 7) confirm the significant effect for each variable: expectation of continuity [$F(2, 54) = 8.749$, $p < 0.01$; $\eta^2 = 0.245$; power $= 0.962$] and desire for future dealing [$F(2, 54) = 5.941$, $p < 0.01$; $\eta^2 = 0.180$; power $= 0.860$]. Examinations of the means and the pairwise comparisons in Table 8 indicate that expectation of continuity in electronic reverse auction ($M = 4.479$) was significantly lower (1.382 points) than in both the e-mail negotiation ($M = 5.467$) and the face-to-face

Table 4 ANOVA results (DV: satisfaction with dealing)

Source	Sum of squares	d.f.	Mean squares	F	p
Trust	26.315	1	26.315	37.913	0.000
Mechanism	21.569	2	10.784	15.538	0.000
Trust × Mechanism	4.764	2	2.382	3.432	0.039

Table 5 Bonferroni pairwise comparisons between mechanisms (DV: satisfaction with dealing)

Mech (I)	Mech (J)	Mean Difference (I–J)	Std. Err	p
F2F	E-mail	0.331	0.279	0.720
F2F	eRA	1.357	0.262	0.000
E-mail	eRA	1.026	0.254	0.000

Table 6 MANOVA multivariate tests results (DVs: expectation of continuity and desire for future dealing)

Source		Value	F	d.f.	p
Trust	Pillai's trace	0.409	18.313	2	0.000
Mechanism	Pillai's trace	0.251	3.874	4	0.006
Trust × Mechanism	Pillai's trace	0.067	0.936	4	0.446

Table 7 MANOVA univariate tests results (DVs: expectation of continuity and desire for future dealing)

Source	DV	Sum of squares	d.f.	Mean squares	F	p
Trust	Expectation of continuity	44.377	1	44.377	36.622	0.000
	Desire for future dealings					
Mechanism	Expectation of continuity	19.016	1	19.016	17.902	0.000
	Desire for future dealings					
Trust × Mechanism	Expectation of continuity	21.204	2	10.602	8.749	0.001
	Desire for future dealings					

Table 8 Bonferroni pairwise comparisons between mechanisms (DVs: expectation of continuity and desire for future dealing)

Measure	Mech (I)	Mech (J)	Mean difference (I–J)	Std. Error	p
Expectation of continuity	F2F	E-mail	0.394	0.368	0.866
	F2F	eRA	1.382*	0.349	0.001
	E-mail	eRA	0.988*	0.338	0.015
Desire for future dealing	F2F	E-mail	0.417	0.345	0.696
	F2F	eRA	1.097*	0.327	0.004
	E-mail	eRA	0.681	0.317	0.108

* Significant at the corresponding p value

negotiation (0.988 points lower; $M = 5.861$). A significant difference in desire for future dealing of 1.097 points was observed between electronic reverse auction ($M = 5.042$) and face-to-face negotiation ($M = 6.139$). Examination of the means also showed differences in the predicted direction between face-to-face and e-mail negotiations; however, the effects were small and not statistically significant.

In summary, our data fully supported Hypothesis 1b and partially supported Hypothesis 1c (for satisfaction with dealing and expectation of continuity); however, data fail to confirm Hypothesis 1a.

Hypothesis 2 predicted that the level of pre-existing trust between the buyer and the supplier would have a positive impact on the relational outcomes, that is, higher pre-existing trust would be linked to higher satisfaction with dealing, higher expectation of continuity and higher desire for future dealing than would low levels of pre-existing trust. The analysis of the univariate test on satisfaction with dealing (Table 4) indicated a significant main effect of the trust manipulation ([$F(1, 55) = 37.913$, $p < 0.001$; $\eta^2 = 0.408$; power $= 1.000$], namely a significant difference in satisfaction with dealing existed between the high trust ($M = 6.166$) and low trust conditions ($M = 4.833$). Significant multivariate effects [Pillai's trace $= 0.409$, $p < 0.001$, $\eta^2 = 0.409$, power $= 1.000$] (Table 6) and univariate effects (Table 7) were observed for expectation of continuity ([$F(1, 54) = 36.622$, $p < 0.001$; $\eta^2 = 0.404$; power $= 1.000$]) and desire for future dealing ([$F(1, 54) = 17.902$, $p < 0.001$; $\eta^2 = 0.249$; power $= .986$]. Expectations of continuity were significantly higher in high trust ($M = 6.139$) than in the low trust condition ($M = 4.399$), as well as for desire for future dealing ($M = 6.204$ in high trust condition vs. $M = 5.065$ low trust condition). Therefore, Hypothesis 2 was fully supported.

Hypothesis 3a and 3b predicted that the face-to-face negotiation would result in higher relational outcomes than would the e-mail negotiation and electronic reverse auction, respectively, in the low trust condition, but not in the high trust condition. Conversely, Hypothesis 3c predicted that the differences in relational outcomes between e-mail negotiation and electronic reverse auction would be significant at both high and low levels of trust. Analysis of the between-subjects tests showed a significant two-way interaction effect between trust and sourcing mechanism on satisfaction with dealing [$F(2, 55) = 3.342$, $p < 0.05$; $\eta^2 = 0.111$; power $= 0.621$] (Table 4); conversely the trust $\times$ sourcing mechanism interaction effect on expectation of continuity and desire for future dealing was not significant [Pillai's trace $= 0.067$ ns]. In order to test which mechanisms differ on satisfaction with dealing at the two different levels of trust, two parallel ANOVA models were designed to analyze the effect of mechanism at each level of trust separately. Pairwise comparisons of the means indicate that sourcing mechanism had a much larger effect on satisfaction with dealing under low than high trust conditions (Table 9). Specifically, under conditions of low pre-existing trust, satisfaction was significantly lower after the electronic reverse auction than it was after either the e-mail or face-to-face negotiation. In contrast, there were no significant differences in satisfaction across sourcing mechanisms under conditions of high trust. These findings provide support for Hypothesis 3b, but not for Hypothesis 3a or Hypothesis 3c.

Hypotheses 4a, b , and c predicted a significant differences for high and low trust on satisfaction with dealing in face-to-face negotiation as well as in the e-mail

Table 9 Bonferroni pairwise comparisons between mechanisms in parallel models high trust versus low trust (DV: satisfaction with dealing)

Trust condition	Mech (I)	Mech (J)	Mean difference (I–J)	Std. Error	p
High trust	F2F	E-mail	0.162	0.400	1.000
	F2F	eRA	0.715	0.370	0.193
	E-mail	eRA	0.553	0.357	0.401
Low trust	F2F	E-mail	0.500	0.387	0.620
	F2F	eRA	2.000*	0.371	0.000
	E-mail	eRA	1.500*	0.360	0.001

Table 10 Bonferroni pairwise comparisons between trust levels in the three mechanism (DV: satisfaction with dealing)

Trust condition	Trust (I)	Trust (J)	Mean difference (I–J)	Std. Error	p
F2F	High	Low	0.792	0.277	0.012
E-mail	High	Low	1.130	0.323	0.003
eRA	High	Low	2.077	0.417	0.000

negotiation and electronic reverse auction conditions. These predictions find support in empirical data. In fact, we found significant differences in satisfaction with dealing between the high and low trust conditions for all three mechanisms (Table 10): electronic reverse auctions showed the largest difference at 2.077 points between high ($M = 5.744$) and low trust ($M = 3.667$). Smaller though significant differences were observed for e-mail negotiation and face-to-face negotiation.

6.5 Results for Trust and Perception of Opportunism

Hypotheses 5a, b, and c predicted that the richest sourcing mechanism (face-to-face negotiation) would increase perceived trust most from the pre-test to post-test phase, followed by e-mail negotiations and electronic reverse auctions. These hypotheses were tested with a multivariate mixed-design analysis of variance in which: the *test phase* (pre-test versus post-test) is the within-subject factor, and *sourcing mechanism* and *trust condition* are the between-subject factors.

As shown in Table 11, within-subjects multivariate tests indicate a significant main effect of time [Pillai's trace $= 0.362$, $p < 0.001$, $\eta^2 = 0.362$, power $= 0.999$] on perceived trust and opportunism. Univariate tests and means analyses indicate that the average level of perceived trust increased significantly from pre-test ($M = 3.843$) to post-test ($M = 4.474$), whereas the average level of perceived opportunism decreased significantly from pre-test ($M = 4.016$) to post-test ($M = 3.287$).

In addition, analyses showed a significant two-way interaction between time and sourcing mechanism [Pillai's trace $= 0.230$, $p < 0.01$, $\eta^2 = 0.115$, power $= 0.859$],

Table 11 Repeated measures MANOVA multivariate tests results (DV: trust and perception of opportunism)

Source		Value	F	d.f.	p
Within-subjects effects					
Time	Pillai's trace	0.362	15.319[a]	2.000	0.000
Time × Trust	Pillai's trace	0.598	40.221[a]	2.000	0.000
Time × Mechanism	Pillai's trace	0.230	3.581	4.000	0.009
Time × Trust × Mechanism	Pillai's trace	0.020	0.272	4.000	0.895
Between-subjects effects					
Trust	Pillai's trace	0.895	2.312	2.000	0.000
Mechanism	Pillai's trace	0.185	2.804	4.000	0.029
Trust × Mechanism	Pillai's trace	0.111	1.612	4.000	0.176

indicating that the change in perceived trust and opportunism from pre-test to post-test differed across the three mechanisms. Pairwise comparisons for the three mechanisms indicate a significant pre-test to post-test change in perceived trust and opportunism for the face-to-face and e-mail negotiations, but not for the electronic reverse auction. The means, which are presented graphically in Figs. 3 and 4, showed significant increases in perceived trust from pre-test ($M = 3.781$) to post-test ($M = 4.931$) and significant decreases in opportunism from pre-test ($M = 3.822$) to post test ($M = 2.671$) when face-to-face negotiation was used. The changes were smaller though significant when e-mail negotiations were used (trust increased by 0.597 points from pre-test to post-test and opportunism decreased by 0.989 points). In the electronic reverse auction condition, levels of trust and opportunism did not change significantly from pre-test to post-test. Therefore, Hypothesis 5a and Hypothesis 5b are supported, whereas Hypothesis 5c is not.

Hypothesis 6 predicted that the trust condition would significantly affect the development of perceived trust and opportunism, and Hypothesis 6a and 6b offered competing predictions concerning the direction of this effect. Analyses showed a significant two-way interaction between time and trust condition [Pillai's trace $= 0.598$, $p < 0.001$, $\eta^2 = 0.598$, power $= 1.000$], supporting Hypothesis 6.

Fig. 3 Variation of trust across time and mechanism

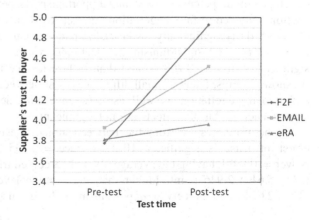

Fig. 4 Variation of
perceived opportunism across
time and mechanism

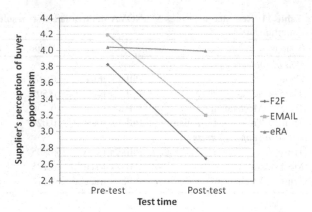

Results show a significant positive effect of time on perceived trust when the pre-existing trust condition was manipulated to be low: in fact, perceived trust increased by 1.768 points from pre-test ($M = 1.792$) to post-test ($M = 3.560$). Conversely, when the pre-existing trust condition was manipulated to be high, the results showed a small but significant decrease in subsequent levels of perceived trust (0.506 points) from pre-test ($M = 5.893$) to post-test ($M = 5.387$). With regard to perceived opportunism, a significant downward trend was observed when the pre-existing trust condition was low, namely a decrease of 1.500 points from pre-test ($M = 5.586$) to post-test ($M = 4.085$). When the pre-existing trust condition was high, perceived opportunism did not change significantly from pre-test to post-test. Thus, Hypothesis 6a was not supported, whereas Hypothesis 6b was partially supported (for trust but not for perceived opportunism).

Additional analyses related to the predictions discussed so far may provide additional insight into understanding how pre-existing trust levels and sourcing mechanisms influence relational outcomes. For example, between-subjects multivariate tests indicate a significant main effect of sourcing mechanism [Pillai's trace $= 0.185$, $p < 0.05$, $\eta^2 = 0.093$, power $= 0.751$]. Bonferroni pairwise comparisons across mechanisms indicate that the only significant difference in the development of perceived trust and opportunism is between electronic reverse auction and face-to-face negotiation. In fact, trust in electronic reverse auction ($M = 3.891$) is significantly lower (0.465 points) than in face-to-face negotiation ($M = 4.356$), and opportunism in electronic reverse auction ($M = 4.014$) is significantly higher (0.768 points) than in face-to-face negotiation ($M = 3.247$). As shown in Figs. 5 and 6, small differences exist between face-to-face negotiation and e-mail negotiation as well as between e-mail negotiation and electronic reverse auction, but the effects are not significant.

Furthermore, results from the between-subjects tests showed that the main effect of trust is significant [Pillai's trace $= 0.895$, $p < 0.001$, $\eta^2 = 0.895$, power $= 0.999$], having the average level of perceived trust in high trust condition ($M = 5.640$) 2.946 points higher than the average level in low trust condition ($M = 2.676$) and perceived opportunism in high trust condition ($M = 2.468$)

Fig. 5 Trust across
mechanisms

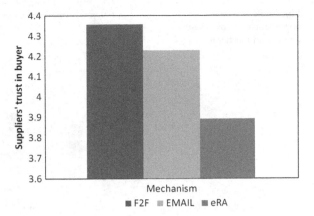

Mechanism

■ F2F   ■ EMAIL   ■ eRA

Fig. 6 Perceived
opportunism across
mechanisms

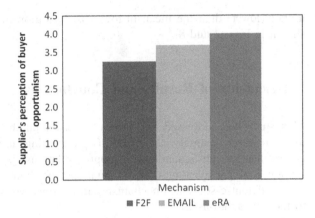

Mechanism

■ F2F   ■ EMAIL   ■ eRA

2.368 points lower than in low trust condition ($M = 4.836$). These results also confirm that the manipulation of the levels of trust (high versus. low) between buyer and supplier was effective. Besides, the effect of trust does not vary across mechanisms: the mean of perceived trust (opportunism) in high trust condition is

Fig. 7 Trust across trust
levels and mechanisms

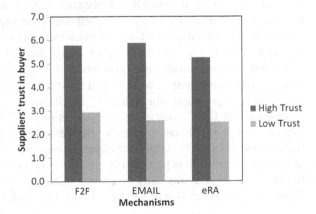

■ High Trust
■ Low Trust

Mechanisms

Fig. 8 Perceived
opportunism across trust
levels and mechanisms

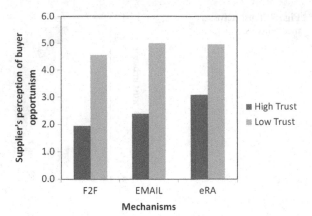

higher (lower) than the mean in low trust condition in all three mechanisms, as shown in Figs. 7 and 8.

7 Discussion of Results and Conclusions

This study has addressed key issues in recent purchasing and supply management research, namely the role of pre-existing interorganizational trust as determinant of subsequent buyer–supplier relationships. We have focused on the influence of pre-existing trust on relationships, both alone and in conjunction with different e-sourcing mechanisms in contrast to the use of traditional face-to-face negotiations.

A 3 × 2 design of experiments was used to compare three sourcing mechanisms (face-to-face negotiation, e-mail negotiation, and electronic reverse auction) and two pre-existing levels of trust between the buyer and the supplier (high and low). Behavioral laboratory experiments were selected as methodology that would enable us to draw strong conclusions about the causal nature of our predicted relationships. Using a controlled laboratory setting, in which participants performed the role of either a buyer or a supplier of transportation service, we were able to isolate the impact of the independent variables from other extraneous factors (e.g., differences in the contract terms) that could affect results.

Multivariate and univariate analyses of variance, which were performed on data from the suppliers' sample, confirm that a considerable difference existed between face-to-face negotiation (the "richest" mechanism in the study) and electronic reverse auction (the "leanest" mechanism in the study) in terms of their relative impact on the relational outcomes. E-reverse auctions caused lower supplier satisfaction in dealing with the buyer, compared to face-to-face and e-mail negotiation; this result is particularly significant in the low trust scenario compared to the high trust scenario, thus suggesting that electronic reverse auctions may be ill-suited to repair trust when pre-existing relationships were poor. Interestingly, we found that e-mail negotiations, which are often neglected for strategic trans-actions, achieved relational outcomes comparable to face-to-face negotiations.

The obtained results provide useful insights for selecting sourcing mechanisms which are appropriate for a desired or given type of business relationship. As this research links different mechanisms available for sourcing practices to their relational performance, results can help companies to trade-off transactional against relational outcomes, thereby selecting the mechanism best fitting the given trust scenario and the desired type of interorganizational relationship.

7.1 Limitations

Several limitations of our study should be recognized. One issue is that we manipulated the level of pre-existing trust using verbal instructions. Although the manipulation check indicated significant differences in perceived supplier trust consistent with the manipulation, it is possible that an actual history of strong or weak relationships might produce different outcomes in actual sourcing contexts. For example, it may be more difficult to overcome a long history of involving low trust through a single e-mail or face-to-face negotiation. It would be helpful for future researchers to explore this in field settings.

Second, the terms of the transaction were strictly limited in both the negotiation and the electronic reverse auction conditions. Although we designed the transaction to approximate a complex transaction including three attributes (price, delivery interval, and reliability), it may be that contracts involving more attributes or requiring a longer time-frame for negotiation than was permitted here would lead to somewhat different effects on relational outcomes. Future researchers might vary the number of attributes in future laboratory studies as one way to explore this possibility.

Finally, our predictions focused solely on relational outcomes and not on the actual terms of the transaction, which has been a central focus in many prior studies (e.g., Carter and Stevens 2007). We did not test whether pre-existing trust or sourcing mechanisms led to differences in the total value of the deal reached in this transaction, and it is possible that objective differences in terms may affect relational outcomes over time.

7.2 Practical Implications

Our study suggests several practical implications that may be of benefit to buyers and suppliers. One key finding is that in the context of high pre-existing trust, electronic reverse auctions may not necessarily damage existing relationships. Note that this contradicts some prior findings, such as that of Tassabehji et al. (2006) who reported the electronic reverse auctions increased skepticism among suppliers. Several factors may explain this discrepancy: for example our reverse auction participants "competed" against two other suppliers. Participating in auctions that involve a larger number of suppliers has been linked to increased perceived opportunism (Carter and Stevens 2007), so it is possible that using e-reverse auctions with trusting supplier under less competitive conditions mitigates losses in trust.

A second implications is that both face-to-face and e-mail negotiations were linked with increases in perceived trust and reductions in perceived opportunism. This suggests that richer sourcing mechanisms may be used as trust-building or trust-repairing strategies. Buyers who have a history of poor relationships with suppliers thus may consider using these sourcing mechanisms to build trust, and postpone using electronic reverse auctions.

Appendix. Measures

Trust

1. This company keeps the promises it makes to my company.
2. This company is not always honest with my company (reverse coded item).
3. My company believes the information that this company provides us with.
4. When making important decisions, this company considers my company's welfare as well as its own.
5. My company finds it necessary to be cautious with this company (reverse coded item).
6. This company is genuinely concerned that our business succeeds.
7. My company trusts this company to keep our best interests in mind.
8. This company is trustworthy.

Perception of opportunism

1. In future interactions, I believe this company would be unwilling to accept responsibility for its mistakes.
2. In future interactions, I believe this company would provide us with false information.
3. In future interactions, I believe that this company would try to "nickel and dime" us.

Satisfaction with dealing

1. Dealing with this company benefits your company.
2. My company is satisfied with the dealings with this company.
3. This company is a good company to do business with.

Expectation of continuity

1. I expect to continue working with this company on a long-term basis.

2. The relationship with this company will last far into the future.

Desire for future dealings

1. Based on your experience in this negotiation, to what degree are you willing to have future dealings with this company?

References

E. Ambrose, D. Marshall, B. Fynes, D. Lynch, Communication media selection in buyer-supplier relationships. Int. J. Oper. Prod. Manag. **28**(4), 36–379 (2008)

AMR. Are we moving from buyer and seller to collaborators? SCM report, American Manufacturing Research Inc. (1998)

S.S. Andaleeb, An experimental investigation of satisfaction and commitment in marketing channels: the role of trust and dependence. J. Retail **72**(1), 77–93 (1996)

D. Andersson, A. Norrman, Procurement of logistics services—a minutes work or a multi-year project. Eur. J. Purchas. Supply Manag. **8**, 3–14 (2002)

S. Beall, C. Carter, P.L. Carter, T. Germer, T. Hendrick, S. Jap, L. Kaufmann, D. Maciejewski, R. Monczka, K. Petersen, The role of reverse auctions in strategic sourcing. Technical report, CAPS Research (2003)

E. Bendoly, R. Croson, P. Goncalves, K. Schultz, Bodies of knowledge for research in behavioral operations. Prod. Oper. Manag. **19**(4), 434–452 (2008)

E. Bendoly, K. Donohue, K. Schultz, Behavior in operations management: assessing recent findings and revisiting old assumptions. J. Oper. Manag. **24**, 737–752 (2006)

W.C. Benton, M. Maloni, The influence of power driven buyer/seller relationships on supply chain satisfaction. J. Oper. Manag. **23**, 1–22 (2005)

M. Bichler, An experimental analysis of multi-attribute auctions. Decis. Support Syst. **29**, 249–268 (2000)

M. Bichler, G.E. Kersten, S. Strecker, Towards a structured design of electronic negotiations. Group Decis. Negot. **12**(4), 1035–1054 (2003)

D.T. Campbell, J.C. Stanley, *Experimental and quasi-experimental designs for research* (Haughton-Miflin, Boston, 1963)

M.C.J. Caniels, E.M. van Raaij, Do all suppliers dislike electronic reverse auctions? J. Purchas. Supply Manag. **15**(12), 23 (2009)

C.R. Carter, C.K. Stevens, Electronic reverse auction configuration and its impact on buyer price and supplier perceptions of opportunism: a laboratory experiment. J. Oper. Manag. **25**, 1035–1054 (2007)

C. Chen-Ritzo, T. Harrison, A.M. Kwasnica, D.J. Thomas, Better, faster, cheaper: an experimental analysis of a multiattribute reverse auction mechanism with restricted information feedback. Manage. Sci. **51**(12), 1753–1762 (2005)

J.A. Colquitt, Publishing laboratory research in AMJ: a question of when, not if. Acad. Manag. J. **51**(4), 616–620 (2008)

T.D. Cook, D.T. Campbell, *Quasi-experimentation: designs and analysis issues for field settings* (Rand McNaly College Publishing Company, Chicago, 1979)

R. Croson, K. Donohue, Behavioral causes of the bullwhip effect and the observed value of inventory information. Manage. Sci. **52**(3), 323–336 (2006)

P.M. Doney, J.P. Cannon, An examination of the nature of trust in buyer-supplier relationships. J. Mark. **61**(2), 35–51 (1997)

M.L. Emiliani, D.J. Stec, Aerospace parts suppliers' reaction to online reverse auctions. Supply Chain Manag. Int. J. **9**(2), 139–153 (2004)

T.F. Gattiker, X. Huang, J.L. Schwarz, Negotiation, email, and internet reverse auctions: how sourcing mechanisms deployed by buyers affect suppliers' trust. J. Oper. Manag. 44(3), 53–75 (2007)

W.T.Y. Ghijseen, J. Semeijn, S. Ernstson, Supplier satisfaction and commitment: the role of influence strategies and supplier development. J. Purchas. Supply Manag. 16(1), 17–26 (2009)

F. Gino, G. Pisano, Towards a theory of behavioral operations. Manuf. Serv. Oper. Manag. 10, 676–691 (2008)

J. Hair, R. Anderson, R. Tatham, W. Black, Multivariate Data Analysis (Prentice Hall, Upper Saddle River, 1998)

R. Handfield, S. Straight, What sourcing channel is right for you. Supply Chain Manag. Rev. 7(4), 62–68 (2003)

X. Huang, T.F. Gattiker, J.L. Schwarz, Interpersonal trust formation during the supplier selection process: the role of communication channel. J. Supply Chain Manag. 44(3), 53–75 (2008)

S.D. Jap, An exploratory study of the introduction of on line reverse auctions. J. Mark. 67, 96–107 (2003)

S.D. Jap, The impact of online reverse auction design on buyer-supplier relationships. J. Mark. 71, 146–159 (2007)

D.A. Johnston, D.M. McCutcheon, F.I. Stuart, H. Kerwood, Effects of supplier trust on performance of cooperative supplier relationships. J. Oper. Manag. 22(1), 23–38 (2004)

L. Kaufmann, C. Carter, Deciding on the mode of negotiation: to auction or not to auction electronically. J. Supply Chain Manag. 40(2), 15–26 (2004)

A. Losch, J.S. Lambert, E-reverse auctions revisited: an analysis of context, buyer-supplier relations and information behavior. J. Supply Chain Manag. 43(4), 47–63 (2007)

R.M. Morgan, S.D. Hunt, The commitment trust-theory of relationship marketing. J. Mark. 58(3), 20–38 (1994)

M. Morris, C.R. Carter, Relationship marketing and supplier logistics performance: an extension of the key mediating variables model. J. Supply Chain Manag. 41, 32–43 (2004)

C.E. Naquin, G.D. Paulson, Online bargaining and interpersonal trust. J. Appl. Psychol. 88(1), 113–120 (2003)

R.L. Oliver, P.V. Balakrishnan, B. Barry, Outcome satisfaction in negotiation: a test of expectancy disconfirmation. Organ. Behav. Hum. Decis. Process. 60, 252–275 (1994)

P.M. Panayides, Y.H. Venus Lun, The impact of trust on innovativeness and supply chain performance. Int. J. Prod. Econ. 122, 35–46 (2009)

J.M. Purdy, P. Nye, P.V. Balakrishnan, The impact of communication media on negotiation outcomes. Int. J. Conflict Manag. 11, 162–187 (2000)

M.J. Sanzo, M.L. Santos, R. Vazquez, L.I. Alvarez, The effect of market orientation on buyer-seller relationship satisfaction. Ind. Mark. Manage. 32, 327–343 (2003)

K. Seliviaridis, M. Spring, Third party logistics: a literature review and research agenda. Int. J. Logist. Manag. 18(1), 125–150 (2007)

D. Simchi-Levi, P. Kaminsky, E. Simchi-Levi, Designing and managing the supply chain: concepts, strategies and case studies (McGraw-Hill International, Singapore, 2000)

T.P. Stank, T.J. Goldsby, S.K. Vickery, Effect of service supplier performance on satisfaction and loyalty of store managers in the fast food industry. J. Oper. Manag. 17(4), 429–447 (1999)

R. Tassabehji, W.A. Taylor, R. Beach, A. Wood, Reverse e-auctions and supplier-buyer relationships: an exploratory study. Int. J. Oper. Prod. Manag. 26(2), 166–184 (2006)

Y. Tian, F. Lai, F. Daniel, An examination of the nature of trust in logistics outsourcing relationship: empirical evidence from China. Indus. Manag. Data Syst. 108(3), 346–367 (2008)

S.K. Vickery, C. Droge, T.P. Stank, T.J. Goldsby, R.E. Markland, The performance implications of media richness in a business-to-business service environment: direct versus indirect effects. Manage. Sci. 50(8), 1106–1119 (2004)

J.G. Wacker, A definition of theory: research guidelines for different theory-building research methods in operations management. J. Oper. Manag. 16(3), 361–385 (1998)

O.E. Williamson, The economic institutions of capitalism. (The Free Press, New York,1985)

C. Young-Ybarra, M. Wiersema, Strategic flexibility in information technology alliances: the influence of transaction cost economics and social exchange theory. Organ. Sci. **10**(4), 439–459 (1999)

A. Zaheer, B. McEvily, V. Perrone, Does trust matter? exploring the effects of interorganizational and interpersonal trust on performance. Organ. Sci. **9**(2), 141–159 (1998)

Chapter 4
Lean Supply Chains: A Behavioral Perspective: Examples from Packaging Supply Chains in the FMCG Sector

Pauline Found

Abstract This chapter discusses the behavioral perspective of lean supply chains which consist of two elements: high-performance operational lean/JIT and high-performance relationship management that are each characterized by distinguishing concepts and features. The extent and successful implementation of operational lean/JIT is contingent upon the product variables; production volume, product standardization, and demand variability. Whereas the high-performance relationship management elements are dependent on the length of relationship, characteristics of the organization, and the policies and practices that are perceived as trustworthy and equitable by both partners. Trust is an important element in both the operations and relationships of lean supply chain management and a maturity path exists where a successful operational lean transformation is highly dependent on the existence of a strong supply relationship based on mutual trust and equity. The conclusions and implications of this study are that a "one-size-fits-all" approach is inappropriate to supply chain design. A contingency approach, that considers all the variables associated with product and organizational factors, is necessary to design an effective and sustainable lean supply chain.

1 Introduction

A unique production system emerged in Japanese manufacturing in the late 1950s that, by the late 1990s, heralded a change throughout the operations management and supply chain literature. This unique and high-performance production/supply

P. Found (✉)
Lean Enterprise Research Centre Cardiff Business School, Cardiff Business Technology Centre, Senghennydd Road Cardiff CF, Wales 24 4AY, UK
e-mail: FoundPA1@cardiff.ac.uk

I. Giannoccaro (ed.), *Behavioral Issues in Operations Management*,
DOI: 10.1007/978-1-4471-4878-4_4, © Springer-Verlag London 2013

system was pioneered and refined by the Toyota Motor Company (Womack et al. 1990).

In the aftermath of the first oil shock in 1973, when market conditions changed, many other Japanese manufacturing companies adopted this approach (Monden 1983), and, by 1977, interest was spreading to the West. Initially the system was known as the *Toyota Production System*. In October 1980, Andersen Consulting organized one of the first Japanese productivity seminars at the Ford Motor Company world headquarters and the term Just-in-Time (JIT) caught the imagination of the Western world (Hall 1981 cited in Harmon and Peterson 1990). The concept behind this high-performance system was a basic logic of "producing the necessary units in the necessary quantities at the necessary time" (Monden 1983, p. 4).

The term *Lean* was later coined by John Krafcik (1988) to describe the philosophy underlying the Toyota Production System (Standard and Davis 1999, p. 49) and was developed further by Womack, Jones, and Roos in their book *The Machine that Changed the World* (1990) which sought to explain the productivity differences between the Japanese and Western automakers. Lean production "encompassed a new production paradigm, a corporate strategy model and an integration model....The lean producer assumes the benefits of just-in-time, total quality, total employee involvement, etc., and builds a global strategy on that basis" (Lamming 1993, p. 18). Fundamental to the JIT and lean philosophy is that manufacturers and suppliers need to work together to provide defect-free components at the right time and in the right quantity.

The terms JIT and Lean are often used interchangeably and are not well defined. JIT is described as a philosophy, a set of techniques, and a method of planning and control (Rich 1999), although Slack et al. (1998) interpret lean as the philosophy and JIT as the management techniques and control methods. While recognizing that reliable supplies are essential to the functioning of JIT production in a lean enterprise, the subject of JIT or lean purchasing has received far less attention in the literature. Burton (1988), Naumann and Reck (1982), and Willis and Huston (1990) estimate that purchased materials and services account for 50–80% of the total cost of manufactured product and it is also estimated that suppliers account for 30 % of the quality problems and 80 % of the lead time (Waters-Fuller 1995). This provides not only considerable scope for improvement and cost reduction, but, with little or no safety stock, could determine the success, or failure, of the lean implementation (Manoochehri 1984). Ansari and Mondarress (1988) argue that JIT efficiency is primarily achieved through complete support, collaboration, and cooperation of suppliers.

According to the literature trust, communication, successful collaboration, good decision making, and business performance are positively correlated. Droge et al. (2004) state that, "Firms recognize that the performance of suppliers' products and the performance of their own products are inextricably linked. Supplier partnering moves beyond supplier development activities and treats suppliers as a strategic collaborator. Supplier partnering approach seeks to bring all the participants in the product lifecycle into the process early on so that each can provide input into the other's processes. Thus, partnership often entails early supplier involvement in product design and/or access to superior supplier technological capabilities (see Narasimhan and Das 1999). Close integration ensures unity of effort and responsiveness.

The integrated supply chain is in many ways synonymous with the Lean Enterprise described by Womack and Jones (1994, p. 93–94; 1996) as "a group of individuals, functions, and legally separate but operationally synchronized companies. The notion of the value stream defines the lean enterprise. The group's mission is collectively to analyze and focus on the value stream so that it does everything involved in supplying a good or service (from development and production to sales and maintenance) in a way that provides maximum value to the customer".

According to Christopher (1992, p. 18) the focus of supply chain management is on cooperation and trust so that the whole can be greater than the sum of the parts. Therefore, supply chain management (SCM) is the management of *relationships* in order to provide a more profitable outcome for all parties of the network. At the heart, the supply chain is a commercial relationship that is affected and influenced by a number of factors. Cox (2003) argues that Japanese supply practices tend to be characterized by high levels of buyer dominance over supplicant suppliers. He further contends that the sourcing options and relationship management approach is contingent upon the demand and supply circumstances of the interchange; horses for courses (Cox 2003). Other authors agree with this and suggest that factors that are critical to the buyer–supplier relationship such as trust, commitment, cooperation, compliance, conflict, and conflict resolution are strongly influenced by power (Brown et al. 1995; Maloni and Benton 2000). Rich and Hines (1997) reject the argument that Japanese companies have achieved greater supply chain integration and operational benefits as a result of the power imbalance and this is supported by Ouchi (1981) who considers that the use of exertion would not achieve the necessary investments by the supplier.

Gonzalez-Benito and Spring (2000) examined purchasing in the Spanish auto components industry and describe two elements of JIT (Lean Supply) purchasing; an operational and a complementary component. Complementary practices include relationship, involvement, and quality elements. The features of the complementary practices are described elsewhere in the literature as *relationship contracting* (Sako 1992; Dore 1987), *partnership sourcing* (Carlisle and Parker 1989; Ellram 1991; Macbeth and Ferguson 1994); *co-makership* (Merli 1991); *lean supply model* (Lamming 1993), and *network sourcing* (Hines 1994, 1996). The features of these include: long-term alliances with few suppliers, single or dual sourcing for each product group and mutuality in problem solving and benefit sharing, open-book costing and transactions. The operational features include frequent deliveries of small lots controlled by the use of Kanbans, or shared inventory management, to reduce stocks and lead time this is facilitated by information that is transferred effortlessly and transparently through IT systems, such as Electronic Data Interchange (EDI), that makes information sharing accessible to all potential supply partners.

The commercial relationship between buyer and seller seeks to minimize the transaction costs, the added costs that are generated by performing a transaction, for example, search costs to find a supplier, costs of generating a purchase order, drafting and negotiating a contract, managing and monitoring the process flow, holding inventories, delivery and transportation, servicing and maintaining ongoing agreements, communication and establishing relationships. In fact, transaction costs encompass virtually everything besides true production costs and exist in every exchange relation (Sako 1992). The optimizing or minimizing of transaction costs is considered to be an important driver in the development of an organizational structure (Williamson 1985), Galvin and Fauske 2000 claim "transaction costs shape the organizational behavior and structure" and Leffler et al. (1991) state "contracting parties will choose the organizational and contractual forms which minimize the costs of transacting".

2 Lean Supply Chain Management

Lean thinking does not start or end with the production process. Within an organization it requires a fundamental change from discrete departments, all jealously guarding their own empires, roles, ideas, information, and direct reports to a new form of *'collaborative'* organization. Communication barriers have to be broken down and information made transparent and easily available. This requires a shift toward a process view of cross-functional teams dedicated to problem solving and driving out waste to enhance value and optimize the value stream. The latter concept concerns the end-to-end processes that deliver value to customers. These include all the sequences of operations as much as it concerns the optimization of supplier and logistics channels to market. The goal of the *lean* supply chain manager is to find a solution, a combination of outsourced and insourced products and services, that economizes on the sum of production, transaction, and management costs. One approach is to choose an organizational form that minimizes cost; the other is to develop cooperative trading relationships based on trust and developing a strategic network that is economically viable.

A collaborative supply chain could simply mean that two or more independent companies work jointly to plan to execute supply chain operations with greater success than when acting in isolation. (Simatupang and Sridharan 2002). Alternatively, collaboration is described as a particular degree of relationship among (supply) chain members as a means to share risks and rewards that result in higher business performance than would be achieved by the firms individually. (Lambert et al. 1999). Recently the sustainability of collaborative supply chains has been questioned (Barratt 2004; Fawcett and Magnan 2002; Sabath and Fontanella 2002). The problems range from difficulties in implementation (Sabath and Fontanella 2002), overreliance on technology (McCarthy and Golocic 2002) and lack of trust between trading partners (Ireland and Bruce 2000; Barratt 2004). Barratt (2004) considers that there is a greater need for understanding of the basic

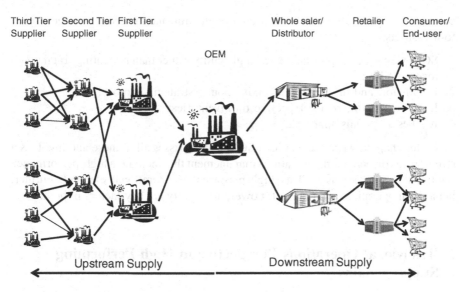

Fig. 1 The shape of "ordered" supply chains

elements of collaboration, particularly the integration of the relevant strategic, cultural, and implementation elements and argues that internal collaboration has the potential to enable internal integration and overcome functional myopia yet has proven elusive as organizations have pursued external collaboration to the detriment of internal issues. The type and level of collaboration depends on the scope of the collaboration: vertical, horizontal, or both (Simatupang and Sridharan 2002).

Upstream and downstream activities (Fig. 1) are therefore part of the lean enterprise and collaboration and communication with suppliers and customers is essential if the product is to flow seamlessly from raw materials to customer. According to Womack and Jones (1996, p. 241), "Even Toyota, the leanest organization in the world, has not yet fully succeeded in creating lean enterprises from raw materials to finished product". Although their first- and second-tier suppliers (direct and indirect levels) operate their production facilities in accordance with the Toyota Production System, their third-tier suppliers are inconsistent. The upstream raw materials suppliers have, so far, resisted Toyota's attempts to streamline their operations and are still firmly stuck in batch production. Raw material suppliers of steel, aluminum, glass, and resins account for 42 % of Toyota's manufacturing cost (Hines 1997) so the real challenge for Toyota is to convince these to change their thinking and behaviors.

Oliver and Webber are described by Svensson (2001) as the founders of the concept SCM. They conducted a study of organizations in the US, Japan, and Western Europe and concluded that traditional approaches to integrate logistics channels failed. "We needed a new perspective and, following from it, a new approach: supply-chain management" (Oliver and Webber 1982, p. 64). They contend that SCM differs from traditional production and materials management in four respects.

1. SCM views the supply chain as a single entity rather than relegating fragmented responsibility;
2. It calls for, and in the end, depends upon, a strategic approach;
3. It provides a different perspective on inventories;
4. It takes a systems approach.

In this chapter, we argue that lean SCM encompasses all of these and involves a strong focus on behavioral operations management that includes a high-performance relationship element as well as high-performance operational element. Further, behavioral operations around trust, power, and equity are key factors in both.

3 Behavioral Operations Perspective in High Performing Supply Chains

Traditional purchasing and supply management practices have been described as adversarial and *arms-length* where buyers and suppliers have negotiated on price. Within these purchasing environments there have been little need, or desire, to develop close relationships. The unit of currency was the purchase order that resulted from a single transaction. Traditional supply relationships and supply chain partnerships are compared in Table 1 that shows that short-term contracts are replaced by long-term alliances with few suppliers. In these environments relationships take on a strategic importance where trust, commitment, and power influence the strength and quality of the trading arrangement.

Two forms of contractual relationships are described by Sako (1992, pp. 9–29) as Arms-length Contractual Relation (ACR) and Obligational Contractual Relation (OCR) which represent the two ends of a multi-dimensional spectrum. ACR is typified by discrete economic transactions where the account is settled at the conclusion of the transaction. Neither party is obliged to continue the relationship nor are they controlled by the other. All dealings are conducted at arm's length and if unforeseen problems arise they are settled by legal or other rules. In contrast

Table 1 Adapted from Stuart 1993

Traditional supply relationships	Lean supply chain partnerships
Focus on cost for supplier selection	Multiple criteria for supplier selection
Short-term *contracts* for suppliers	Long-term *alliances* with suppliers
Large supplier base	Few suppliers
Proprietary information	Shared information
Suppliers are perceived as part of the problem	Suppliers are involved in finding *solutions* to problems

OCR involves a contract that is embedded in mutual trust and is characterized by a high level of interdependence.

The ACR-OCR framework illustrated in Sako 1992, p. 16 takes a system view of the factors influencing the inter-firm relationships to describe buyer–supplier transactions. The framework takes the view that no economic transactions take place in a vacuum but are influenced by a complex socio-economic environment. This section discusses some of the relationship theories and concepts that influence the inter-company and inter-personal relationships within an integrated supply network.

3.1 Power

The literature suggests that factors that are critical to the buyer–supplier relationship such as trust, commitment, cooperation, compliance, conflict, and conflict resolution are strongly influenced by power (Brown et al. 1995; Maloni and Benton 2000). Lukes (1994) defines power as "the ability of actor A to make actor B act in a manner it might not have done". Depending on how the dominant party chooses to use the power-dependency relationship, purchasing strategies can be described as competitive, cooperative, and command.

Cox has been at the forefront of the debate about the role of power in supply chain relationships. He examines the buyer–supplier power relationships within the supply chain and questions the assertions of the lean community of the win–win and trusting long-term relationships of *lean supply* (Lamming 1993; Lamming et al. 2001); *network sourcing* (Hines 1994); and *partnering* (Macbeth and Ferguson 1994). According to Cox (1997, 2002, 2004) Japanese supply practices tend to be characterized by high levels of buyer dominance over supplicant suppliers (Cox 2004, p. 348) and Toyota's structural dominance approach (Cox 1999, p. 172). He argues that, although the *agile* school agrees in principle with long-term collaborative relationships, they point out that "the high volume and highly standardized demand and the supply circumstances in the car industry are not replicated in all other types of industries. In many industries—fashion goods, construction, publishing, for example—demand and supply vary significantly making *lean* approaches to sourcing very difficult" (Cox 2004, p. 348). He argues that the sourcing options and relationship management approach is contingent upon the demand and supply circumstances of the interchange; horses for courses (Cox 2003).

According to Cox (2003) the effect of buyer–supplier power is one of both facilitation and constraint. The desired outcome of the dominant party will be facilitated while those of the dependent party will be constrained. Maloni and Benton (2000) showed empirically, the importance of power within the supply chain. Their findings are summarized below:

- Power plays a significant role in the supply chain, and the different sources of power have contrasting effects on inter-firm relationships in the chain. Thus, both the power source and the power target must be able to recognize the presence of power, and then reconcile supply chain strategy for power influences.
- Exploitation of the supply chain by the power partner may lead to dissension and under performance, thus hurting the power holder. Likewise, a judicious use of power may serve to benefit the power holder.
- Influences of power on the buyer–supplier relationship and subsequent effects of this relationship upon supply chain performance expose the potential of power as a tool to promote integration of the chain and empower higher levels of performance. This performance benefit incites the power holders to take a second look at their positioning of power within supply chain strategy and urges a more conscious, considerate use of power.

Rich and Hines (1997) reject the argument that Japanese companies have achieved greater supply chain integration and operational benefits as a result of the power imbalance and propose a three pillar methodology as a general framework. They claim that policy deployment, cross-functional management, and supplier integration is generally applicable to most organizations and cite examples from distribution (RS Components) to FMCG (Proctor and Gamble).

However, according to Benton and Maloni (2005) "it may be argued that a firm with significant power might not find it necessary to establish the win–win alliance since it can achieve its own profitability and effectiveness through control of its suppliers (dependents). In other words, firms with the bargaining power have little if any reason to yield control or to withhold exercise of such power. In seeking their own profitability and success, the dominant firms may be better off pursuing their own individual supply chain agendas, submitting to a joint planning partnership only as much as the balance of power dictates".

3.2 Trust

Transaction costs are considered to be reduced in a relationship with high levels of trust (Williamson 1985). Jarillo (1993) focuses on the reduction of transaction costs in a network where entrepreneurs invest in building mutual trust. Trust has a role in reducing uncertainty and risk in an economic transaction thereby reducing transaction costs (Sako 1992, p. 37). In her book, *Prices, Quality, and Trust* Mari Sako (1992, p. 38–39) defines three categories of trust:

1. *Contractual trust*. The moral duty of both partners to execute their obligations. Suppliers are trusted to produce the required quantity of goods at the specified time and buyers are trusted to pay for these within the time agreed. The contract may be written as in a purchase order, or contract, but may also be verbal.

Contractual trust relies on the keeping of promises and covers explicit and implicit agreements.

2. *Competence trust.* The expectation that the trading partner is technically competent to perform the exchange. On behalf of the supplier this is to supply the product or service to the required specification and for the buyer to competently specify and make the transaction and payment.
3. *Goodwill trust.* The trust that is expressed as a willingness to do more than expected.

Both ACR and OCR relationships rely on *contractual trust* and *competence trust* but *goodwill trust* exists only in OCR relationships. According to Sako "What distinguishes 'goodwill trust' from 'contractual trust' is the expectation in the former case that trading partners are committed to take initiatives (or exercise discretion) to exploit new opportunities over and above what was explicitly promised" (p. 39).

Contractual trust and *competence trust* can be gained by screening and audits but *goodwill trust* is contextual and only gained within a relationship and develops over time and through shared experiences. Therefore, the investment costs are very high and a long-term perspective is required to recoup the investment. Fawcett et al. (2004, p. 20) think that trust is still not clearly understood by managers in the west but argue that a lack of trust is "the greatest obstacle to advanced supply chain collaboration". They further describe four dimensions of supply chain trust:

1. The performance dimension—trust depends on consistently doing what you say you will do.
2. The information sharing dimension—trust requires open and clear information and vice versa open and clear information requires trust.
3. The behavioral dimension—sharing of risks and rewards and the willingness to invest in supply chain partners capabilities.
4. The personal dimension—trust is personal and developed through one-on-one meetings and customer and supplier visits made by cross-functional teams.

According to Fawcett et al. (2004, p. 24) Honda's Teruyuki Marou said "Suppliers do not trust purchasing because purchasing means cost, but they must trust you. Suppliers must develop confidence in you." These views are supported by other writers. Bowersox et al. (2000) state "effective information sharing is heavily dependent on trust beginning within the firm and ultimately extending to supply chain partners". Ellram and Cooper (1993) and Gardner and Cooper (1993) consider that if information is shared openly then opportunistic behaviors are reduced. Beccerra and Gupta (1999, p. 197) agree and state "in low-trust relationships, people protect themselves by sharing less information and taking more conservative actions". They also consider the personal dimensions and acknowledge that "business occurs among people who have biases, cultures, attitudes, experiences, and interact with each other through time. A certain level of trust will grow with each relationship" (Beccerra and Gupta 1999, p. 198). The use, or rather misuse, of coercive power on the other hand diminishes trust.

Trust can be both the antecedent and consequence of asset specificity (Ik-Whan and Suh 2004) but the relationships are very complex. Asset specificity affects trust and transaction costs. Where asset specificity is high the risk of opportunistic behavior is greater and trust is lower resulting in higher transactions costs. However, the converse is also true; where trust is high the willingness to invest in specific assets is also higher; but as these assets increase, the dependency on the supply chain partner also increases. The possibility of opportunistic behavior is much more damaging when one party has assets highly specific to the relationship (Beccerra and Gupta 1999. Ik-Whan and Suh (2004) propose that trust leads to commitment while Fynes et al. (2005) present empirical evidence that communication has a positive effect on trust that, in turn influences commitment and adaptation (transaction specific investments) that positively correlate with quality and cost.

3.3 Equity

The supply chain network consists of links and nodes that are individual firms. The network chain, or web, is only as strong as its weakest link. Satisfaction has a key role in strengthening the relationship. "Thus, a manufacturer cannot be responsive without *satisfied* suppliers, and the benefits of such a relationship cannot be transferred to the end customer unless the distributors align with this manufacturer's strategy as well. At the same time, a manufacturer cannot produce quality products without pushing quality responsibility upstream to its suppliers. SCM involves the strategic process of coordination of firms within the supply chain to competitively deliver a product or service to the ultimate customer" (Benton and Maloni 2005, p. 2). They define supplier satisfaction in the supply chain as "a feeling of *equity* with the supply chain relationship no matter what power imbalances exists between the buyer–seller dyad."

The notion of "equity" is associated with justice and fairness. The individual fundamentally believes that they are being treated fairly in comparison to what they see others receiving. Adams (1963) advanced the proposition that we each, on acting to satisfy our needs, assess the equity or fairness of the outcome we perceive. Adams equity theory can be applied to the manufacturer—supplier dyads to motivate partners to work for the optimization of the whole chain.

Torrington et al. (2002) describing Adams' Equity Theory state that "we are concerned that rewards or outputs equate to our inputs and that these are fair when compared to the rewards being given to others." They argue that low trust relationships exist where people feel they are not being treated fairly. This would suggest that where powerful business customers exert their power unfairly the suppliers perceived equitable rewards are reduced and trust is compromised.

The concepts of power and trust do much to create a model of relationship management but these concepts are not very informative unless motivations are investigated and therefore the concept of supplier equity in the relationship is seen as the final aspect upon which to ground this study. Having outlined the background theory underpinning supply chain management, this chapter will return to the features of modern lean businesses, the use of power, trust, and equity in the design of supply chains that support high performing businesses.

4 Evolution of the Lean Supply Model

4.1 Supply Chain Management in the West

Purchasing practices at Ford's Highland Park plant consisted of dual sourcing and competitive tendering with prices held for six to twelve months (Lamming 1993 citing Sorenson 1956). Henry Ford distrusted his suppliers and proceeded to vertically integrate. This became an obsession until he owned everything. The Model T was built with Ford-owned glass and steel made from Ford-owned ore, coal, and timber but huge capital investment bought inflexibility and high fixed costs. The complex was so inflexible that even small changes came at huge costs (Lamming 1993 citing Chandler 1964). However, the component supply model for the first decades of mass manufacturing was classic vertical integration. Although General Motors followed this approach they were accidentally introduced to subcontracting and in 1921 under Sloan decided not to operate in the component market but rely more on outside firms.

The European component suppliers originated in the highly skilled craft industry and although, US style mass manufacturing ideas were popular in which many small specialized firms remained. However, some large assemblers approached high levels of vertical integration. Following the Second World War, the political climate in Europe resulted in a number of companies coming under public ownership and financed wholly, or partly, from government sources. This gave rise to strong national identity characteristics and company-specific issues dominated (Lamming, op. cit. p 14).

A major transformation occurred in the post-war years in the customer base for automobiles; they wanted variety, and also at this time tariffs were removed allowing the Europeans to develop a steady market for their small cars in the US. This gave them the opportunities to gain economies of scale and to be able to compete with the US giants. With competition came the demand for sophisticated and innovative components that were more successfully produced from independent component suppliers; although vertical integration with fiercely adversarial out-sourcing remained the norm for US and European automotive suppliers for many years to come (Lamming op. cit. p 16).

4.2 Supply Chain Management in Japan

By contrast to Western style supply chain management, the Japanese had evolved from a very different financial model. The first era of Japanese industrialization saw family-owned holding companies, *zaibatsu*, controlling industrial empires that consisted of a large company in each major sector steel, shipbuilding, construction, insurance, finance (Womack et al. 1990, p. 193). Banks were included in the holding company and finance for investment came directly from the bank. These *zaibatsu* were disbanded, along with their assets, by the Americans during the occupation of Japan following World War II.

Nishigushi (1987, 1994) describes the emergence of *strategic dualism* in 1920s as the foundation for Japanese subcontracting. Dualism, a strategy through which an assembler tries to outsource certain items while maintaining the manufacture of key components, was the strategy that Chrysler traditionally followed until the late 1980s. Relationships between assembler and supplier are extensively *arm's length* (Nishigushi 1987, p. 2). Subcontracting developed rapidly from 1930s onwards in response to sudden surges in demand. During the period 1931–1939 Toyota developed its supply base by buying parts directly from the US companies, disassembling them and seeking local firms to copy the parts (Cusumano 1985, p. 64). There were no long-term relationships with suppliers at this time and innovation and quality were poor (Cusumano 1985, p. 66). As subcontracting became more widespread purchasing became more important. Both Toyota and Nissan developed their purchasing departments during this time. In 1939 around 66 % of the manufacturing costs, excluding raw materials, of a Toyota motor vehicle was attributable to purchased parts. Purchased components were divided into three types (Nishigushi 1994, p. 37):

- General purchasing
- Special purchasing
- Specialty factory purchasing

General purchasing is for items that require no specialized manufacture and can be bought from many suppliers, who can easily be switched. Special purchasing and specialty purchasing are with suppliers that have expertise and require close ties both, financial and/or capital; signaling a move toward asset specific resources. This arrangement was strategic in intent and set in Toyota's internal rules.

Supplier associations (*kyoryokukai*) were a product of the government's wartime program of organizing the subcontractors into channeled groups (*keiretsui*) (Nishigushi 1994, p. 39). In contrast to the financial holding of a *zaibatsui, keiretsui* members are held together by cross-locking equity structures and the system was glued together by a sense of reciprocal obligation (Womack et al. 1990, p. 194–195). Toyota's *kyoryokuka* was formed in 1943 between Toyota and twenty key subcontractors. Strategic dualism remained the basis of *keiretsui* supplier relationships until the 1960s with prime contractors beating down prices until the government intervened to prevent unfair practices. During this time the automotive sector were pioneering more harmonious and goodwill relationships but these were not apparent in the post-war Japanese electronics industries that were neither benevolent nor trusting.

Considerable changes emerged in the 1960s that lay the foundations for Japanese models of *relationship contracting* (Sako 1992; Dore 1987), *partnership* (Carlisle and Parker 1989; Ellram 1991; Macbeth and Ferguson 1994); *Co-makership* (Merli 1991); *lean supply model* (Lamming 1993); and *network sourcing* (Hines 1994, 1996). Network sourcing is described as "a model derived from the observation of best practice buyer–supplier relationships from around the world, but particularly from Japan" (Hines 1996, p. 19). Hines (1994, 1996, p. 8) identified ten characteristics that defined the Japanese Network Sourcing model:

1. A tiered supply structure with heavy reliance on small firms;
2. A small number of direct suppliers with individual part numbers sourced from one supplier, but within a competitive dual sourcing environment;
3. High degrees of asset specificity among suppliers and risk sharing between customer and supplier alike;
4. A maximum 'buy' strategy by each company within the semi-permanent supplier network, but a maximum 'make' strategy within these trusted network;,
5. A high degree of bilateral design employing the skills and knowledge of both customers and suppliers alike;
6. A high degree of supplier innovation in both new products and processes;
7. Close, long-term relations between network members involving a high level of trust, openness, and profit sharing;
8. The use of rigorous supplier grading systems increasingly giving way to supplier self-certification;
9. A high level of supplier coordination by the customer company at each level of the tiered supply structure;
10. A significant effort made by customers at each level individually to develop their supplier.

The network sourcing model recognizes the reorganizing of subcontracting into tiers that moved the supply structure from semi-arm's length to a systematic *clustered control* (Nishiguchi 1994, p. 122) based on a pyramid structure. The firms at the apex of the pyramid buy complete assemblies and system components from a concentrated, clustered base of first-tier subcontractors, who buy specialized parts from second-tier suppliers, who buy discrete parts or labor from third-tier subcontractors, etc. Nishigushi (1987) recognizing that the *keiretsu* is not a closed system reorganized the pyramids into an interlocking form that he called the Alps structure.

The shift from discrete purchasing to complex asset-specific industrial contracts required different means of pricing and value analysis (VA) techniques developed by General Electric's purchasing department were adopted and became widely used in Japan from 1960s. Detailed cost breakdowns of value-added components paved the way to rational price determination rather than negotiating price downstream. Suppliers and subcontractors investigated ways of reducing costs by joint improvements, sharing the benefits and buyer–supplier profit sharing rules were developed and traditional unilateral price determination shifted to bilateral price agreements (Nishigushi 1996, p. 125). Along with this shift in price negotiations was the move toward participation in new production development and innovation. Nishigushi (1996, p. 125) describes it as "the logic of contractual relations moved from exploitation to collaborative manufacturing".

Ansari and Modarress (1988) argue that lean supply chain efficiency is primarily achieved through complete support and cooperation of suppliers. They list the following activities as major components of lean supply purchasing.

1. Small purchase lot sizes, delivered in exact quantities.
2. Few suppliers, ideally one per component or family of parts
3. Supplier selection and evaluation based on quality and delivery performance as well as price, rather than solely a price decision
4. No incoming quality inspection
5. Looser design specifications giving the supplier more freedom in meeting specifications
6. No annual rebidding compared to traditional annual tendering
7. Standard containers
8. Reduced and less formal paperwork.

Other authors (Manoochehri 1984; Freeland, 1991; Schonberger and Gilbert 1983) have, in addition to these, included other practices.

9. Deliveries synchronized to buyer's production schedule
10. Geographically close suppliers
11. Improved data exchange

Other studies of lean supply (JIT) and purchasing in the Spanish auto-components industry (Gonzalez-Benito et al. 2000; Gonzalez-Benito and Spring 2000; Gonzalez-Benito and Suarez-Gonzalez 2001 and Gonzalez-Benito 2002) found three factors that need to be taken into account when designing a high-performance supply chain: product variables (the characteristics of the exchanged product or service); the buyer and seller organizational variables and variables associated with the marketing environment.

In a study of Mexican manufacturing plants, Lawrence and Lewis (1996) reported that quality, customer service, and productivity were higher where JIT logistics and supplier involvement practices were noted. This supported by Fawcett and Birou (1993) who argued that there is a direct relationship between JIT purchasing and reported financial benefits: reduced administrative, inspection and inventory costs, as well as other benefits such as: quality, productivity, and improved scheduling.

The literature highlighted many common implemented, or technical, features which support high performance—the lean supply features. These design issues have been used to compile the model. Table 2

Relationship theories are applied to a study of the supply chain for premium printed packaging to two FMCG companies where the packaging is an integral part of the product, used for protection, information and differentiation of the product

Table 2 The concepts of high-performance relationship management identified in the literature

Concept	Features	Author(s)
High-performance relationship management		
Trust	Involvement, commitment, information sharing, and transparency	Sako(1992), Dore (1987), Hines(1994), Nishiguchi (1994), Lamming (1993), Ellram and Hendrick (1995), Fawcett (2004), Chu and Fang (2006), Johnston et al. (2004)
Cost transparency	Better cost control; open-book costing; selfinvoicing	Merli (1991), Lamming (1993), Lamming et al.(2000)
Dependency	Dedicated assets; shared destiny; colocation	Ansari and Modarress (1988), Schonberger and Gilbert (1983), Merli (1991),Lamming (1993), Hines (1994), Gonzalez-Benito and Spring (2000)
High-performance operational JIT/Lean Purchasing		
Kanbans with suppliers VMI	Reduced inventories; improved material availability; frequent deliveries; small lots; standardized containers	Ansari and Modarress (1988), Chen et al. (2005), Chyr et al. (1990), DeToni and Zamolo (2005), Disney and Towill (2003), Dong and Xu (2002)
Time compression (Reduced lead-time)	Reduced batch sizes; reduced set up and changeover times	Fine (1998), Ansari and Modarress (1988), Schonberger and Gilbert (1983), Rich (2002), Gonzalez-Benito (2002)
EDI	Integrated IT systems; production schedules; and stock visibility	Ansari and Modarress (1988), Schonberger and Gilbert (1983), Merli (1991), Ellram and Hendrick (1995), Waller(1991), Gonzalez-Benito and Spring (2000)

from the competition where, because of the uniqueness of the packaging, asset specificity is high and trust is a key element in the supply relationship that influences both the relational and the operational elements of lean supply.

The supply of printed packaging to FMCG receives scant attention in the lean/JIT literature, restricted mainly to flexibility of in-line label printers, yet the reliable provision of high-quality printed packaging, whether this is in the form of folding cartons, decorated tins, boxes, bottles or tubes is essential to maintain the flow of FMCG manufacturing in the consumer-packaged goods sector. This chapter, which builds partly on the work of Gonzalez-Benito and Spring (2000), seeks to address this and presents the findings of a study that investigated the relationships and supply chains of two major FMCG companies with their printed packaging suppliers.

5 Research Approach

This chapter reports the findings from a case study of packaging supply chains in the FMCG sector. The two cases selected represented users of high quality printed packaging designed for use in the high-end FMCG market. Clearly observations from only a few organizations are not likely to be representative of the entire industry. However, Hartley (1994) counters this argument by observing that statistical generalizations might be out of date by the time they are interpreted, whereas a description of the processes might be valuable.

The unit of analysis was the packaging purchasing process carried out within each organization. This avoids the problem of ambiguity by analyzing the process, and the managers perceptions of the process, rather than the overall performance of the purchasing functions within the organizations. Eight purchasing managers from the two case firms were surveyed and account managers, production managers and packaging technologists from six packaging companies in three sectors participated in semi-structured interviews and site visits.

6 Case Studies

Fusion (a coded name) was established in Europe over two hundred years ago to produce a range of paper products in the FMCG market. The product has changed little over the years but the lifestyle marketing of the product and the target consumer market has changed. At the height of production there were five factories throughout Europe, with two in the UK. Subsequent market decline has closed most of these, leaving one in the UK and other one in mainland Europe.

The production involves preparing the paper products and packing them in various formats and styles for end-user consumption through retailers. Fusion is sold by a range of retail customers range from large multi-national supermarkets to small independent shops. Fusion is the brand leader for these products and they operate in a cost-conscious, price sensitive market where on-shelf availability and quality reliability are key competitive drivers.

Phobos (a coded name) was established in the UK in 1900 and has served the UK market continuously since this time. They have successfully established markets in Europe and the Far East over the past two decades. Phobos produces high-end, luxury consumer products in a competitive environment. The packaging serves to protect the product for consumption and to differentiate it from the competition. The products are packed in a variety of formats including gift packs and comprise folded cartons, decorated tins, and printed laminates and films.

Both organizations have a long and varied history in supplying FMCG to the European market and both have, at some time in their past, been separately vertically integrated, owning their own in-house printers and controlling their own packaging supply chains. Both now purchase their packaging from UK and European printers and converters. Packaging materials represent over 20 % of the final product cost and is a significant element of the product. The criticality of the product is such that any component of the packaging that is sub-standard, or not available, means that the product cannot be shipped and, in most cases, cannot complete manufacture.

The research compared the results found in the case studies with the published literature on the characteristics of lean supply chains (Manoochehri, 1984; Ansari and Modarress 1988; Freeland 1991; Schonberger and Gilbert 1983). However, the packaging supply chain, in contrast to other studies, is characterized by highly developed suppliers, often more technologically advanced than the customers. In addition, the suppliers produce for a variety of FMCG customers and are exposed to a number of advanced practices and ideas, often ahead of more traditional organizations, so some of the concepts, such as supplier development and training, applies more to supply chains where the advanced organization is the downstream producer.

7 Research Findings and Reflections

The findings support the conclusions of Gonzalez-Benito (2002) that there are two components of lean supply; an operational component and a relational component (Fig. 2). In this study, the packaging supply chain was found to be relatively well developed in the relational components that are influenced by organizational characteristics but poorly developed in the operational elements that are influenced by the product characteristics such as volume, variety, specificity, and economic value.

These findings are consistent with the findings of Gonzalez-Benito et al. (2000) that operational and complementary practices can be implemented separately. The operational practices are inherent in a lean environment but the complementary practices that depend on trust and cooperation between supply partners are appropriate to all manufacturing environments. They further prove that operational

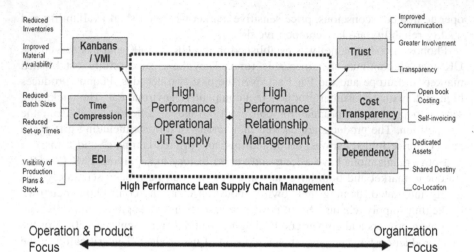

Fig. 2 Model of high-performance lean supply chain management

practices cannot be fully implemented without the presence of the complementary practices, suggesting a maturity pathway.

The concepts of high-performance relationship management accepted from the literature review are:

- Trust
- Cost transparency
- Dependency

7.1 Trust

The features of a trusting relationship are:

- Improved communication and information sharing
- Transparency and openness of information
- Involvement in new products and processes

Trust reduces transaction costs due to absence of opportunistic behaviors. Power and equity affect trust but power comes in several forms:

Coercive power	based on the ability to punish
Reward power	based on the ability to reward
Legitimate power	comes with the formal position or title
Referent power	-based respect or charisma
Expert power	that comes from having expertise in a particular area

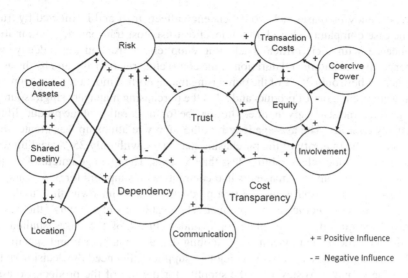

Fig. 3 Interactions of variables in high-performance relationship elements

The use, or abuse, of coercive or reward power will achieve compliance but will erode trust and encourage suppliers to behave opportunistically if they feel a perceived loss of equity. The interactions are shown in Fig. 3.

7.2 Cost Transparency

Cost transparency facilitates open book costing and negotiations based on costs rather than price. In an environment of cost transparency self-invoicing is encouraged. Cost transparency depends on trust and cannot exist without trust.

7.3 Dependency

The features of dependency are:

- Dedicated assets
- Shared destiny
- Co-location
- Dedicated assets can be:

    - Physical assets—machines, tools or site
    - Human assets—skills, capabilities or labor resources

Dependency increases risk; so dependency affects trust and is affected by trust.

The case companies shared a high level of trust, cost transparency, and mutual dependence with their critical packaging suppliers. Trust and dependency was influenced by the nature and duration of the contractual relationship and by the asset specificity where only 25 % of the purchasing managers had more than two suppliers for any given category of product and all of the purchasing managers single-sourced each product line at any given time. This may be for the duration of the product life or the supply contract. The average length of the supply relationship was greater than 10 years for the majority of the purchasing managers, with only 25 % dealing with suppliers where the relationship is less than 5 years. The average length of supply contract is 1 year but two managers had contracts with suppliers for 3–5 years.

As packaging is usually unique to a product, the supplier owns the "tools" or "plates" that are specific to that product. The customer only "owns" the design engraved, or etched, on the plate, or cylinder. In all of the cases studied the supplier holds dedicated machinery or tooling that is product, or brand-specific. In 75 % of cases, the cost of moving to another supplier influenced the decision of the purchasing manager to stay with the supplier for the life of the product. So asset specificity is a dominant factor in printed packaging and is highly influential in supplier selection and choice as supplier switching is unlikely.

This model indicates where the relationship variables of trust, power, and equity positively or negatively impact on each other and/or act to increase or decrease the adoption of the elements of a supply relationship model. For example dedicated assets, shared destiny, and colocation increases risk and dependency yet, while high risk can negatively impact on trust, dependency is strengthened in a trusting relationship. Similarly transaction costs are increased in a high risk environment where trust is low and coercive power is high. This model was used by the purchasing managers and supplier account managers to rank their perceptions during the cross case comparisons and cross data displays.

However, trust is also a very important variable in the successful implementation of the operational features of a lean supply chain. This supports the work of Gonzalez et al. (2000) who consider that operational JIT purchasing cannot be implemented before the complementary practices. Figure 4 illustrates the interactions between the operational variables.

The characteristics of the product such as volume, standardization and demand variability determine the feasibility of operational JIT supply implementation. While none of the packaging suppliers mentioned that they developed Kanbans with any customers, several said that they had produced JIT for large customers where the products were standard and the demand stable. All of the suppliers reported that they managed stocks for some of their customers; this is either vendor managed inventory (VMI) against agreed minimum stock levels, or comanaged (CMI). For some FMCG manufacturers with high volume turnaround or volatile demand, the packaging is held at the customer as consignment stocks and invoiced on crossing the line into production. All of the suppliers in this study reported that they had some customers who self-invoiced, either once the materials had been consumed from stock or on receipt into their own stock system. Many of the suppliers have EDI arrangements

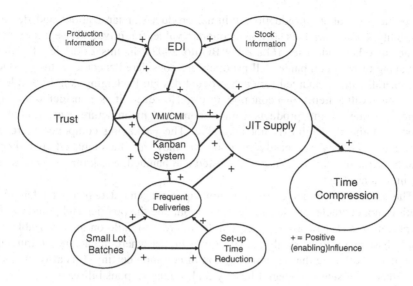

Fig. 4 Interactions between variables of high-performance operational lean supply chains

with some of their customers and generally requirements are transmitted electronically. Some customers give the suppliers visibility of their production schedules but most work from sales forecasts. While not fully adopting lean manufacturing, the main packaging suppliers are actively involved in continuous improvement and various Lean/Six Sigma programs with their customers. Trust appeared to be as significant an element in operational as in relational supply management.

To develop a lean supply model based on reducing inventories and compressing lead-time advocates of lean suggest that a flow, or pull system, is implemented. Central to this would be reducing batch sizes and establishing JIT deliveries synchronized to real customer demand. Inventory models such as VMI or CMI may be part of this solution, as would kanbans or other forms of demand signaling. All of these require a high degree of trust. The manufacturer would need to trust that the supplier could meet the tight delivery schedules with good quality parts or materials. The supplier would need to trust that the customer would provide them with high quality data to manage their own production and to meet the payment terms. Without trust a lean supply could not be established and could not function. In a true high-performance lean supply chain sharing of electronic data or production schedules is possible, but again this can only happen in a relationship of mutual trust.

8 Conclusions

The main research findings are that lean SCM consists of two elements, operational JIT/lean purchasing and high-performance relationship management. The extent and successful implementation in the packaging industry is contingent upon the product variables; production volume, product standardization; and demand variability. Other variables such as economic value, fragility, and specificity have impact in other industries (Gonzalez-Benito 2002). In this example of printed packaging to both companies, all products are considered specific as the products are generally single-sourced for each product design. Packaging is not considered to be high-value items in economic terms. However, it is considered of high strategic value as the products cannot complete manufacture, and cannot be shipped, if the packaging is not available. The packaging components are not fragile in that they are easily broken, but they can be damaged and, hence unsuitable for manufacture, if poorly stored and handled; a risk that is increased if inventories are high.

The success of implementing relationship management depends on length of relationship, characteristics of the organization, and the policies and practices that are perceived as trustworthy and equitable. In order to build on these variables the supply base needs to be small and success is higher where the actors are mutually dependent and where there is a perception of common destiny. It is also enhanced when there is a sense of shared history and mutual responsibility.

The implementation of high-performance relationship management is independent of the industry and can be applied irrespective of the production environment. Conversely, the JIT/lean operational practices depend on the supply relationships that support them and they can only be fully implemented in a JIT/lean operating environment (Fig. 5). This suggests that there is a maturity path toward the Lean Enterprise which starts by developing high-performance relational management that is a necessary precursor to lean production and operational JIT/lean. Trust appears to be a significant element to both relational and operational lean supply chain management.

Undoubtedly a strong customer has power over a smaller, or dependent, supplier, and can force the supplier to comply with customer requirements. However, coercive power does not elicit long-term commitment and affects the level of trust and equity in the relationship. The risk of the suppliers in a coercive power relationship behaving opportunistically is high and the relationship may not survive long term.

Fig. 5 Maturity path from lean supply relationship management to the lean enterprise

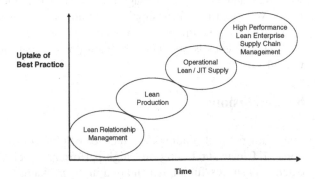

In summary a lean supply chain can be described as a system, whose design is contingent on the product variables, the socio-technical organizational characteristics of all partners and the organizational structures that support them. Behavioral operations of SCM are of crucial importance to developing high-performance lean supply chains.

Finally, to appreciate the benefits, and to proactively seek to improve and implement best practice, requires that the organization understands best practice and has the capacity to learn; this may be contingent on the organization, its structure and its exposure, and openness to new ideas.

References

J.S. Adams, toward an understanding of inequity. J. Abnorm. Soc. Psychol. **67**, 422–436 (1963)

A. Ansari, B. Modarress, JIT purchasing as a quality and productivity centre. Int. J. Prod. Res. **26**(1), 19–26 (1988)

M. Barratt, Understanding the meaning of collaboration in the supply chain. Supply Chain Manage. **9**(1), 30–42 (2004)

M. Beccerra, A.K. Gupta, Trust within the organization: Integrating the trust literature with agency theory and transaction cost economics. Public Administration Quarterly **23**(2), 177–204 (1999)

W.C. Benton, M. Maloni, The influence of power driven buyer/seller relationships on supply chain satisfaction. J. Oper. Manage. **23**, 1–22 (2005)

D.J. Bowersox, D.J. Coss, T.P. Stank, Ten, mega-trends that will revolutionize supply chain logistics. J. Bus. Logist. **21**(2), 1–16 (2000)

J.R. Brown, J.L. Johnson, H.F. Koening, Measuring the sources of marketing channel power: a comparison of alternative approaches. Int. J. Res. Mark. **12**, 333–354 (1995)

T.T. Burton, JIT/repetitive sourcing strategies: tying the knot with your suppliers. Prod. Inv. Manage. J. **29**(4), 38–41 (1988)

J. Carlisle, R. Parker, *Beyond Negotiation; Redeeming Customer-Supplier Relationships, 1989* (Wiley, Chichester, 1989)

F.Y. Chen, T. Wang, Z.X. Tommy, Integrated Inventory Replenishment and Temporal Shipment Consolidation: A Comparison of Quantity-Based and Time-Based Models. Ann. Oper. Res. **135**(1), 197–213 (2005)

M. Christopher, *Logistics and Supply Chain Management: Strategies for Reducing Cost and Improving Service*, 1st edn. (Financial Times Prentice Hall, London, 1992)

S.-Y. Chu, W.-C Fang, Exploring the Relationships of Trust and Commitment in Supply Chain Management. J. Am. Acad. Bus. **9**(1), 224–229 (2006)

F. Chyr, TM. Lin, C. Ho, Comparison between Just-in-Time and EOQ Systems. Eng. Costs Prod. Econ. **18**, 233–240 (1990)

A. Cox, On power, appropriateness and procurement competence. Supply Manage. **2**(20), 24 (1997)

A. Cox, Power, value and supply chain management. Supply Chain Manage. **4**(4), 167–175 (1999)

A. Cox, Supply Chains, Markets and Power: mapping buyer and supplier power regimes. Routledge, London (2002)

A. Cox, Horses for courses. Suppl. Manage. **8**(3), 28 (2003)

A. Cox, The art of the possible: relationship management in power regimes and supply chains. Suppl Chain Manage. **9**(5), 346–356 (2004)

M.A. Cusumano, The Japanese Automobile Industry. Harvard University Press, Cambridge, MA (1985)

A.F. De Toni, E. Zamolo, From traditional replenishment system to vendor-managed inventory: A case study from the electrical appliances sector. Int. J. Prod. Econ. **96**(1), 63–80 (2005)

S.M. Disney, D.R. Towill, The effect of vendor managed inventory (VMI) dynamics on the Bullwhip Effect in supply chains. Int. J. Prod. Econ. **85**(2), 199–214 (2003)

Y. Dong, K. Xu, A supply chain model of vendor managed inventory. Transport. Res. E-Log **38E**(2), 75–81 (2002)

R.P. Dore, *Taking Japan Seriously* (Stanford University Press, Stanford, 1987)

C. Droge, J. Jayaram, K.V. Shawnee, The effects of internal versus external integration practices on time-based performance and overall firm performance. J. Oper. Manage. **22**, 557–573 (2004)

L.M. Ellram, A Managerial Guideline for the Development and Implementation of Purchasing Partnerships. Int. J. Purchas. Mater. Manage. **27**(2), 2–8 (1991)

L.M. Ellram, M.C. Cooper, Characteristics of supply chain management and the implications for purchasing and logistics strategy. Int. J. Logist. Manage. **4**(2), 1–10 (1993)

L.M. Ellram, T.E. Hendrick, Partnering characteristics: A dyadic perspective. J. Bus. Logist. **16**(1), 41–64 (1995)

S.E. Fawcett, L.M. Birou, Just-in-Time sourcing techniques: Current State of Adoption and Performance Benefits. Prod. Inv. Manage. J. **34**(1), 18–24 (1993)

S.E. Fawcett, G.M. Magnan, The rhetoric and reality of supply chain integration. Int. J. Phys. Distrib. Logist. Manage. **32**(5), 339–361 (2002)

S.E. Fawcett, G.M. Magnan, A.J. Williams, Supply Chain Trust is Within your Grasp. Supply Chain Manage. Rev. **8**(2), 20–26 (2004)

C.H. Fine, C.H. 1998. Clockspeed, (Perseus Books, Reading, MA., 1998)

J.R. Freeland, A survey of just in time purchasing practices in the United States. Prod. Invent. Manage. J. **32**(2), 43–49 (1991)

B. Fynes, C.Voss, D.B. Seán, The impact of supply chain relationship dynamics on manufacturing performance. Int. J. Op. Prod. Manage. **25**(1), 6–19 (2005)

P. Galvin, J. Fauske, Transaction Costs and the Structure of Interagency Collaboratives: Bridging Theory and Practice. In Education Policy in the 21st Century, ed. by B.A. Jones (Ablex Publishing Corporation, Stamford, CT, 2000)

J.T. Gardner, M.C. Cooper, Building good business relationships. Int. J. Phys. Distrib. Logist. Manage. **23**(6), 14–26 (1993)

J. Gonzalez-Benito, Effect of the characteristics of the purchased products in JIT purchasing implementation. Int. J. Oper. Prod. Manage. **22**(7/8), 868–886 (2002)

J. Gonzalez-Benito, M. Spring, JIT purchasing in the Spanish auto components industry—implementation patterns and perceived benefits. Int. J. Oper. Prod. Manage. **20**(9), 1038–1052 (2000)

J. Gonzalez-Benito, I. Suarez-Gonzalez, Effect of organizational variables in JIT purchasing implementation. Int. J. Prod. Res. **39**(10), 2231–2249 (2001)

J. Gonzalez-Benito, I. Suarez-Gonzales, M. Spring, Complementaries between JIT purchasing practices: an economic analysis based on transaction costs. Int. J. Prod. Econ. **67**, 279–293 (2000)

R.L. Harmon, L.D. Peterson, Reinventing the Factory Productivity Breakthroughs in Manufacturing today, (Free Press, New York, 1991)

J. Hartley, Case Studies in Organisational Research. In *Qualitative Methods in Organisational Research*, ed. by C. Cassell and G. Symon (London, Sage, 1994)

P. Hines, *Creating World Class Suppliers—Unlocking Mutual Competitive Advantage* (Pitman Publishing, London, 1994)

P. Hines, Purchasing for Lean production: the new strategic agenda. Int. J. Purchas. Mater. Manage. **32**(1), 2–10 (1996)

P. Hines, Toyota Supplier System in Japan and the UK. Proc. 3rd Int. Symp.Logist., Univ of Padua, Padua, 87–96 (1997)

G.K. Ik-Whan, T. Suh, Factors Affecting the Level of Trust and Commitment in Supply Chain Relationships. J. Supply Chain Manage. **40**(2), 4–15 (2004)

R. Ireland, R. Bruce, CPFR only the beginning of collaboration. Suppl. Chain Manage. Rev. Sept/ Oct, 80–88 (2000)

J.C. Jarillo, *Strategic Network; Creating the Borderless Organization,* (Butterworth-Heinmann, Oxford, 1993)

D.A. Johnston, D.M. McCutcheon, F.I. Stuart, H. Kirkwood, Effects of supplier trust on performance of cooperative supplier relationships. J. Oper. Manage. **22**(1), 23–38 (2004)

J.K. Krafcik, European Manufacturing Practice in a World Perspective. Policy Forum Paper, Int. Motor Vehicle Program, MIT (1988)

D.M. Lambert, M.A. Emmelhainz, J.T. Gardner, Building successful logistics partnerships. J. Bus. Logist **20**(1), 165–181 (1999)

R. Lamming, *Beyond Partnership: Strategies for Innovation and Lean Supply,* (Prentice Hall, Hemel Hempstead, Herts, 1993)

R. Lamming, N. Caldwell, W. Phillips, Suppy Chain Transparency. In Understanding Supply Chains, ed. by S. New, R. Westbrook (Oxford University Press, Oxford, 23–42, 2004)

J.J. Lawrence, H.S. Lewis, Understanding the use of just-in-time purchasing in a developing country. The case of Mexico. Int. J. Oper. Prod. Manage. **16**(6), 68–83 (1998)

K.B. Leffler, R.R. Rucker, Transaction Costs and the Efficient Organization of Production: A Study of Harvesting Contracts. J. Pol. Econ. **99**(5), 1060–1087 (1991)

S. Lukes, Power: A Radical View. Macmillan, London (1994)

D. Macbeth, N. Ferguson, *Partnership Sourcing: An Integrated Supply Chain Approach,* (Financial Times Pitman Publishing, London, 1994)

M. Maloni, W.C. Benton, Power influences in the supply chain. J. Bus. Logist. **21**(1), 49–85 (2000)

G.H. Manoochehri, Suppliers and the just in time concept. J. Purchas. Mater. Manage. **20**(4), 16–21 (1984)

S. McCarthy, S. Golocic, Implementing collaborative planning to improve supply chain performance. Int. J. Phys. Distrib. Logist. Manage. **32**(6), 431–454 (2002)

G. Merli, *Co-Makership: The New Supply Strategy for Manufacturers* (Productivity Press, Cambridge, 1991)

Y. Monden, Toyota Production System. Industrial Engineering and Management Press, Norcross, GA (1983)

R. Narasimhan, A. Das, Manufacturing agility and supply chain management practices. Prod. Inv. Manage. J. **40**(1), 4–10 (1999)

E. Naumann, R. Reck, A buyer's bases of power. J. Purchas. Mater. Manage. **18**(4), 8–14 (1982)

T. Nishiguchi, Strategic Industrial Sourcing: The Japanese Advantage. Oxford University Press, New York (1994)

T. Nishiguchi, Competing Systems of Automotive Components Supply: An Examination of the Japanese "Clustered Control" Model and the "Alps" Structure. First Policy Forum IMVP May 1987. Ontario, Canada: MIT. 26 (1987)

R.K. Oliver, M.D. Webber, Supply chain management: logistics catches up with strategy. In Logistics: The Stategic issues, ed. by M. Christopher, (Chapman & Hall, London, 63–75 1982)

W. Ouchi, Organizational paradigms: a commentary on Japanese management and theory Z organizations. Organ. Dyn. **9**(4), 36–44 (1981)

N. Rich, *Total Productive Maintenance: The Lean approach* (Tudor Business Publishing Ltd., Wirral, England, 1999)

N. Rich, P. Hines, Supply-chain management and time-based competition: the role of the supplier association. Int. J. Phys. Distrib. Logist. Manage. **27**(3/4), 210 (1997)

R. Sabath, J. Fontanella, The unfulfilled promise of supply chain collaboration. Supply Chain Manage. Rev. July/August, 24–29 (2002)

M. Sako, *Prices, Quality and Trust: Inter-firm relations in Britain and Japan* (Cambridge University Press, Cambridge, 1992)

R.J. Schonberger, J.P. Gilbert, Just-in-time purchasing: a challenge for US industry. Calif. Manage. Rev. **26**(1), 54–68 (1983)

T.M. Simatupang, R. Sridharan, The collaborative supply chain. Int. J. Logist. Manage. **13**(1), 15–30 (2002)

N. Slack, S. Chambers, R. Johnston, *Operations Management*, 2nd edn. (FT Prentice Hall, Harlow, 1998)

C. Standard, D. Davis, Running today's factory: a proven strategy for lean manufacturing. (Hanser Gardner Publications, Cincinnati, OH., 1999)

L. Stuart, Supplier partnerships; influencing factors and strategic benefits. Int. J. Purchas. Mater. Manage. **29**(4), 22–28 (1993)

G. Svensson, Just-in-time: the reincarnation of past theory and practice. Management Decision **39**(10), 866–879 (2001)

D. Torrington, L. Hall, S. Taylor, Human Resource Management (Financial Times Prentice Hall, Harlow, Middx, 2002)

N. Waters-Fuller, Just-in-time purchasing and supply: a review of the literature. Int. J. Oper. Prod. Manage. **15**(9), 220–236 (1995)

O.E. Williamson, *The Economic Institutions of Capitalism: Firms, Markets, Relational Contracting* (The Free Press, New York, 1985)

T. Willis, C.R. Huston, Vendor requirements and evaluation in a JIT environment. Int. J. Oper. Prod. Manage. **10**(4), 41–50 (1990)

J.P. Womack, D.T. Jones, From lean production to the lean enterprise. Harvard Bus. Rev. **72**(2), 93–103 (1994)

J.P. Womack, D.T. Jones, D. Roos, *The Machine that Changed the World.* (Rawson Associates, New York, 1990)

J.P. Womack, D.T. Jones, *Lean Thinking: Banish Waste and Create Wealth in your Corporation* (Simon & Schuster, London, 1996)

Chapter 5
Supply Chain Integration: A Behavioral Study Using NK Simulation

Ilaria Giannoccaro

Abstract This chapter investigates the benefits of supply chain integration (SCI) by including behavioral factors influencing decision-maker behavior. In fact, the majority of the studies on the topic assume that SCI is pursued by a central planner, which is completely rational and adopts an optimizing behavior. Since the central planner is an individual, such assumptions appear to be too simplistic. In particular, I focus the attention on two main behavioral factors: (1) cognitive capacity and (2) resistance to change. Managers in fact differ in terms of ability to solve a problem. This depends on their ability to conceive alternatives solutions to test and on their cognitive limit in comparing alternatives and recognizing the best one. Resistance to change is an attitude of individuals who are averse to risk and prefer not to modify/try new solutions for fear of poor outcomes, fear of the unknown, or/and fear of realization of faults. To pursue the research aim a simulation analysis using NK fitness landscape is carried out, in which a central planner characterized by four increasing levels of cognitive capability and two levels of resistance to change is engaged to manage a supply chain in an integrated manner. Three types of supply chain structures characterized by increasing complexity are considered and their performances measured. Results show that supply chain performance increases as the level of cognitive capacity improves in all the supply chain structures, while resistance to change decreases supply chain performance even if the effect on performance is quite low.

I. Giannoccaro (✉)
Department of Mechanics, Mathematics, and Management,
Polytechnic University of Bari, Viale Japigia 182 70126 Bari, Italy
e-mail: ilaria.giannoccaro@poliba.it

I. Giannoccaro (ed.), *Behavioral Issues in Operations Management*,
DOI: 10.1007/978-1-4471-4878-4_5, © Springer-Verlag London 2013

1 Introduction

Supply Chain Integration (SCI) advocates that the whole supply chain (SC) from the material suppliers to the end customers has to be managed by adopting an integrated approach. Integration allows the interdependencies existing along the SC to be managed in an efficient and effective way. Interdependencies arise because independent actors are engaged in the activities of the same production process and often have conflicting aims. If the SC is managed by local independent efforts, the interdependencies are neglected and the whole system results are inefficient although each part is optimized (Christopher 1992; Simchi-Levy et al. 2000).

There is an ample body of literature on SCI. It recognizes the role of SCI as a source of competitive advantage (Bowersox et al. 1999) and highlights the attendant benefits in terms of enhanced performances such as cost reduction, improved flexibility, and time saving (Frohlich and Westbrook 2001; Vickery et al. 2003).

Usually, integration is pursued by adopting a centralized decision making approach, namely there is a central planner in the supply chain which controls the entire production process and makes optimal decisions for the system as a whole (Giannoccaro and Pontrandolo 2004). This is the case of industrial practices implementing SCI such as Vendor Managed Inventory and Continuous Replenishment.

In the majority of the studies on SCI the central planner is assumed to be completely rational and informed. Human decision makers however behave very differently from this approximation in practice (Simon 1979). Their personal motives and their behavioral attributes profoundly affect their decision-making process (Mantel et al. 2006). Thus, the assumptions above appears to be too simplistic and the need to introduce behavioral factors in the analysis of operating system mandatory.

In particular, our understanding of SCI and its effect on supply chain performance is lacking in this respect. The success of SCI will depend on the accuracy of our understanding and modeling of human behavior (Bendoly et al. 2006).

Thus, the aim of this paper is to fill this gap by studying SCI taking into account human and behavioral aspects of the decision maker. In particular, we will attempt to assess the effect of two behavioral factors concerning the decision maker (i.e., the level of cognitive capacity and the resistance to change) on supply chain performance, when the supply chain is managed in an integrated manner.

Kaufmann's NK simulation (1993) is used as the research methodology. It is particularly appropriate for studying complex adaptive systems such as supply chains (Choi et al. 2001; Surana et al. 2005; Pathak et al. 2007) and suited to the development of simple theories like the one that we propose here (Davis et al. 2007). Simulation in fact permits an in-depth examination of the behavior of complex 'real world' systems in ways empirical research prohibits, because it can be run many times, allowing the values of the model parameters to be modified in each run and changes to be observed in the outputs (Carley and Grasser 2000; Berends and Romme 1999).

The chapter is organized as follows. First, the theoretical background of the study is presented. Then, the NK model is discussed and the simulation analysis described. Finally, results are illustrated and managerial implications derived.

2 Theoretical Background

2.1 Supply Chain Structures: A Taxonomy

A supply chain is a network of firms collaborating in the development of a product/ service for a final customer. Firms in the supply chain perform one or more activities of the entire production process, starting from the raw material supply to the distribution of the end product/service to the final customer. Thus, the supply chain may be described as a set of activities carried out by the firms.

The supply chain structure is concerned with coordinating all these activities. Various taxonomies regarding the supply chain structure have been provided in the literature. For example, Lambert et al. (1998) describe the supply chain structures in terms of primary and supporting members and types of business links. Stock et al. (2000) classify them based on the degree of geographical dispersion of the operations and on how the supply and the distribution channels are governed. Lin and Shaw (1998) define three types of supply chain structures, using the following variables: number of nodes, number of tiers, type of participants, type of operations, primary business objectives, product differentiation, product architecture, assembly stages, main inventory type, and product life cycle. Ernst and Kamrad (2000) develop a framework, based on the levels of modularization and postponement, which identifies four classes of supply chain structures, namely rigid, flexible, postponed, and modularized.

We characterize the supply chain structure focusing on the interdependencies existing among the activities carried out by supply chain firms, being the existence of such interdependencies requiring integration and making complexity the coordination of supply chains (Simchi Levy et al. 2000).

Interdependencies are primarily due to the division of labor along the supply chain. Indeed, each supply chain partner performs just a few activities of the entire value creation process from the supply of raw materials to the delivery of the final products and/or services to the customer. Thus, the outcome of each supply chain firm depends on the other partners.

Interdependencies differ in number and pattern. This mainly depends on the adopted supply chain strategy. For example, high modularization reduces the number of interdependencies among the constituting components and respective producing activities (Ulrich 1995). This strategy results in a particular pattern of interdependencies which occur not between activities referring to the production of distinct components, but only within a given component (Ethiraj and Levintal 2004). Conversely, low modularization is characterized by a large number of interdependencies among their components. In such a case, each component influences all the others and this makes high the degree of interdependencies among supply chain activities.

Based on the above, we conceptualize the supply chain structure as a set of interdependent activities carried out by supply chain firms. We characterize it on the basis of two taxonomic variables: (1) the number of interdependencies and (2) the pattern of interdependencies.

For our research purpose, we define three main ideal supply chain structures characterized by increasing complexity.

The *linear* structure is the simplest. It is characterized by a small number of interdependencies among activities. Such interdependencies occur only between adjacent activities, i.e., those belonging to the upstream and downstream phases of the production process. This structure is exhibited by serial supply chains modeled as a repeated chain of buyer–supplier relationships. It characterizes distribution supply chains and those implementing the JIT strategy.

The *modular* structure describes supply networks adopting a modular production process, i.e, interdependencies occur only within blocks and not between blocks. Such a structure is shown by the modular production networks observed in the electronics industry in many countries (Sturgeon 2002). The modular structure is more complex than the serial one, because of the higher number of interdependencies.

The *complex* structure characterizes supply chains in which all activities are interconnected among each other. Full interdependencies among firms occur in supply chains designing and producing integral products, where all firms at the same time need to be linked in order to design an effective product. This is the most complex structure because it has the highest number of interdependencies.

2.2 Supply Chain Integration

An ample literature has investigated SCI. Integration means that an entire supply chain is designed and managed as a single entity by a central planner, which plans and controls the whole system so as to optimize the global performance. This is a challenging task because it requires strategic alignment among partners, huge information sharing among firms supported by an ICT infrastructure, strategic partnerships and collaboration between buyers and suppliers, and joint forecasting and planning for controlling supply chain relationships (Lee et al.1997; Morash and Clinton 1998; Frohlich and Westbrook 2001; Zhao et al. 2008).

SCI involve all managerial levels: strategic, tactical, and operational (Stevens 1989) and might be viewed in terms of the functions to integrate, such as marketing, supply, production planning, distribution, and inventory (Ballou et al. 2000). Key business processes that should be linked along the SC include logistics processes (such as demand management, order fulfilment, manufacturing flow management, and procurement) as well as strategic processes (such as customer relationship management, customer service management, new product development, and commercialization (Lambert et al. 1998)).

There are a number of successful SCI practices implemented across a variety of industries. These include the quick response (QR) method, vendor managed inventory (VMI), co-managed inventory (CMI), jointly-managed inventory (JMI), and collaborative planning, forecasting, and replenishment (CPFR).

In the majority of studies on SCI, the central planner is implicitly assumed to be fully rational and adopting an optimizing behavior. The assumption of full rationality implies that the decision maker is able to gather all the information needed, has the cognitive capacity to make the optimal decision by analyzing the collected information, and behaves in the best interest of the system as a whole. However, the central planner is an individual and her/his behavior is much more complex than this approximation.

In the next section we review behavioral studies in the Operations Management field and investigate SCI by relaxing the assumption of full rationality of the decision maker and by introducing human factors describing the decision maker's behavior.

2.3 Behavioral Decision Making in Operations Management

Behavioral Operations Management (BOM) is the discipline that explores the deviations from rationality of the decision makers involved in the management of a system like a process, a project, an organization, or a supply chain (Siemsen 2009). It investigates the impacts of such deviations on performances and analyzes the strategies to improve them (Gino and Pisano 2008; Loch and Wu 2007).

Bendoly et al. (2006) provide a recent review of BOM studies. They identify three modeling assumptions commonly adopted in many different OM contexts and classify studies on the basis of the OM context and the type of assumption examined. Their review highlights the need for further research especially in the supply chain management area.

Mantel et al. (2006) classify BOM studies in three major classes on the basis of the factors influencing human decisions, i.e., personal attributes, risk, and task characteristics. Payne et al. (1993) highlight that motivation, personal relevance, and expertise are critical personal characteristics that can influence the decision-making process. The other most frequently personal attributes of decision makers analyzed in the literature include trustworthiness, cooperativeness, fairness, individualism, mutuality, integrity, ethics, and opportunism (Wathne and Heide 2000; Bendoly et al. 2006; Mantel et al. 2006; Gino and Pisano 2008; Loch and Wu 2007; Hill et al. 2009).

In this study we focus on two factors we believe critical for the management of a SC: (1) the decision maker's cognitive capacity and (2) the decision maker's resistance to change.

As discussed in the previous section, a decision maker pursuing SCI is involved in a coordination problem of managing interdependent activities. The efficiency and efficacy in solving this problem largely depends on the cognitive capacity of the decision maker, which in turn depends on: (1) his/her ability to conceive diverse solutions to be tested and (2) his/her cognitive limit to evaluate the outcomes of the proposed solution and to identify the best one.

Resistance to change is a personal attribute influencing decision maker behavior. It characterizes a decision maker adopting conservative rather than innovative behavior (Kotter 1995). Conservative behavior means that the decision maker will prefer to maintain the common way activities are accomplished, because he/she is averse to risk. Dubrin and Ireland (1993) in fact have highlighted that resistance to change is driven by fear of poor outcomes, fear of the unknown, and fear of realization of faults.

To the best of our knowledge, no study considers the effect of these factors on SCI.

3 Methodology

The NK model was conceived by Kauffman (1993) to study the evolution of biological systems but it has been also successfully adapted to strategic and organization studies (Levinthal 1997; Rivkin and Siggelkow 2003; Rivkin 2000, 2001; Siggelkow and Rivkin 2005).

An NK model consists in a decision problem defined by the number of decisions (N) and the number of interactions among the decisions (K). The decision problem is modeled as a landscape that maps combinations of choices regarding specific decisions (choice configurations) showing their respective payoffs. The solution of the decision problem consists in reaching the highest peak of the landscape, i.e., to identify the specific combinations of choices regarding the given decisions which yield the highest payoff. During the simulation, the system evolves by assuming higher positions on the landscape so as to reach the highest peak. If the highest peak is reached, the system is evolved with success. The more rugged the landscape, the more difficult it is to reach the highest peak and thus to evolve successfully (Kauffman 1993).

In the next section we describe the NK model of the supply chain by defining variables and the landscape. Finally, we explore how the evolution on the landscape is influenced by the behavioral factors described.

3.1 The Model of the Supply Chain Structure

Consider a supply chain made up of a number of firms. Each firm performs a number of activities such as production planning, distribution planning, inventory replenishment, etc. Firms make decisions concerning how to accomplish their activities.

As a consequence, the supply chain can be modeled as set of N decisions made by firms on how to perform each activity. The vector $d = (d_1, d_2, ..., d_N)$ indicates the combinations of choices regarding activities made by the all firms, i.e., the choice configuration. For the sake of simplicity, we will assume that $N = 12$.

Activities in supply chains are in interaction as described in Sect. 2.1. Thus, interactions occur among decisions. K is the number of interactions among decisions. A decision i is interaction with the decision j when the choice concerning j influences the outcome of i. For example, the inventory replenishment decision of the distributor influences the performance of the retailer's inventory replenishment decision; the decision of the retailer to promote a given product will lead to higher benefits if the producer decides to increase the production of that product.

In the model, each decision makes a contribution C_j to the overall supply chain performance, which depends not only on the choice concerning the single decision, but also on how interdependent decisions (K) are resolved.

The specific pattern of interaction among decisions records which decision affects each one. It corresponds to an $N \times N$ matrix where the "x" in the position (i,j) means that the decision j affects decision i.

Fixing N, K, and the influence matrix, a specific supply chain structure is then defined. Coherently to our discussion in Sect. 2.1, three different supply chain structures are considered, whose influence matrixes are depicted in Fig. 1.

The aim of the supply chain is to identify the choice configuration $(d_1, d_2, d_3, ...d_N)$ yielding to the highest supply chain performance. The overall performance of each choice configuration is computed as the average of the N contributions C_j:

$$P(d) = \left[\sum_{j=1}^{N} C_j \right] / N.$$

This decision problem is interpreted as a performance landscape, i.e., the map of the effect of all possible configurations on performances. The supply chain is thus engaged in an adaptive trek across the landscape in search of the highest peak (global peak).

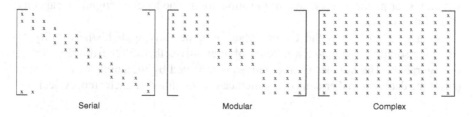

Fig. 1 The influence matrixes of the examined supply chain structures

Pursuing SCI means that a central planner performs just such an adaptive trek across the landscape. He/she is engaged in identifying the choice configuration yielding the highest total payoff $P(d)$. The searching procedure is affected by the behavioral factors characterizing the decision maker's behavior, as we will describe in the next sections. First, we will present the procedure to generate a performance landscape.

3.1.1 The Landscape Generation

The landscape is the map plotting the 2^N configurations and their respective payoffs. It is generated by applying the stochastic procedure described below and models the supply chain structure, because it depends on N, K and the influence matrix.

To generate the landscape, first the specific supply chain structure is selected. Then, the payoff of each configuration is calculated using the formula $P(d) = \left[\sum_{j=1}^{N} C_j \right] / N$, where C_j is the payoff of each single decision d_j.

C_j is drawn at random from a uniform distribution $U(0,1)$. Note that C_j is affected by the choices on the interdependent decisions. Therefore, when $K = 0$, C_j depends only on a single decision, thus C_j assumes the same value in all configurations. When $K = N-1$, as in the complex configuration, C_j depends on how the all other decisions are resolved, thus the C_j differs in all configurations.

This procedure is applied for each supply chain structure, i.e., we generate three types of landscape, i.e., serial, modular, and complex.

3.2 Coding the Behavioral Factors into the Model

3.2.1 Levels of Cognitive Capacity

The decision maker is characterized by one of four increasing levels of cognitive capacity. As described in Sect. 2.3, two factors define the decision maker cognitive capacity: (1) an ability to develop solutions to the problem, (2) a cognitive limit in comparing the alternatives to discover the best one.

Notice that in the model the central planner searches for a new configuration with a higher payoff at each step of the simulation. The level of cognitive capacity affects this search.

In fact, at each step the central planner, coherently with his/her ability to conceive solutions, proposes a number of alternatives that differ from the current configuration. The number of alternatives is modeled by the number of decisions (MD) the decision maker controls, which can be modified at each step. A decision

Table 1 The coding variable of the decision maker cognitive capacity

Decision maker capacities	Coding variables	Options
Ability to conceive alternatives	N. of decisions may differ (MD)	1 versus 3
Cognitive limit to comparison of alternatives	N. of alternatives may be compared (PP)	1 versus All

Table 2 Coding the four levels of the cognitive capacity

Cognitive capacity	Ability	Cognitive limit	MD	PP
Level 1	Low	High	1	1
Level 2	Low	Low	1	12
Level 3	High	High	3	1
Level 4	High	Low	3	298

maker with greater ability can change more decisions at random at the same time, resulting in a greater number of alternatives.

We consider that MD can assume two values: 1 and 3. Only one decision at time is allowed to change when considering a decision maker with low ability; up to 3 decisions change at time for a decision maker with the greatest ability.

Once alternatives are available, the central planner compares the alternative configurations and selects the best one, i.e., the configuration with the highest payoff.

However, due to his/her cognitive limit, the decision maker is able to compare only a subset of alternatives. Such a limit is modeled through the processing power (PP), i.e., the number of total available alternatives that are compared. The higher the PP, the lower the cognitive limit.

Two options for PP are considered: 1 and *all*. PP = 1 means a high cognitive limit, because the decision maker is able to compare only one configuration. Thus, he/she will select one configuration at random among the alternatives. PP = *all* means that all available alternatives are compared and the best one is identified. Thus, the cognitive limit of the decision maker is low.

Table 1 summarizes the code of the two considered factors and the values of the coding variables. All the four possible combinations are considered, resulting in four increasing levels of cognitive capacity.

Table 2 shows for each level of cognitive capacity the values of the coding variables (MD and PP).

3.2.2 Degree of Resistance to Change

Resistance to change is a personal attribute of the decision maker, who prefers to maintain the current way in which the activities are accomplished (status quo configuration), even when alternatives with higher performance for the supply network are available.

We model such an attribute through the probability (p_{RS}) that the decision maker will accept to move in a new configuration with a higher payoff. We consider two options: no $(p_{RS} = 1)$ and high $(p_{RS} = 0.3)$ resistance to change.

The system evolution follows these steps:

1. The central planner conceives MD alternatives;
2. The central planner calculates the payoff of the alternatives;
3. The central planner compares PP alternatives and chooses the alternative with the highest payoff;
4. The central planner adopts the new configuration if this provides a higher payoff, with a probability p_{RC}. Otherwise, status quo is maintained.

4 Simulation Analysis

We designed a plan of experiments consisting of 24 scenarios, resulting from the match between the four levels of the cognitive capacity of the central planner in both cases of high and no resistance to change. In all scenarios $N = 12$ and each landscape was generated 1200 times to guarantee statistical significance to the results. The simulation period was set to 200 steps.

In each scenario we measured the supply chain performance, computed as the system performance $P(d)$ at the end of the simulation, as a portion of the maximum payoff achievable on the landscape. A performance equal to 1 means that the supply chain reached the highest peak on the landscape. Lower values mean that the supply chain during the evolution was trapped in a suboptimal configuration, reaching a payoff lower than the optimum. Results are averaged over the 1200 landscapes.

Table 3 Simulation results

	Serial		Modular		Complex	
	No RC	High RC	No RC	High RC	No RC	High RC
Mean of the performance						
Level 1	0.9451	0.9383	0.9186	0.9132	0.8320	0.8295
Level 2	0.9546	0.9546	0.9377	0.9353	0.8459	0.8457
Level 3	0.9656	0.9304	0.9564	0.9167	0.8782	0.8511
Level 4	0.9898	0.9799	0.9895	0.9784	0.9271	0.9072
Standard deviation						
Level 1	0.0514	0.0544	0.0615	0.0617	0.0600	0.0636
Level 2	0.0479	0.0483	0.0562	0.0570	0.0555	0.0562
Level 3	0.0406	0.0540	0.0421	0.0602	0.0503	0.0571
Level 4	0.0225	0.0311	0.0227	0.0333	0.0402	0.0459

Difference statistically significant with $p < 0.00001$

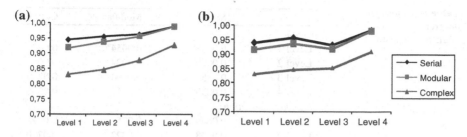

Fig. 2 Performance trends for increasing levels of cognitive capacity

5 Results

The results, reported in Table 3, show that the cognitive capacity of the decision maker and his/her resistance to change affect supply chain performance when supply chain in managed in an integrated manner. Note that there is no scenario where SCI reaches the optimal performance and this is due to the behavioural factors. First, we will discuss the effect of the decision maker's cognitive capacity and then the influence of resistance to change (RC).

Considering a central planner with no resistance to change, increasing the decision maker's cognitive capacity improves supply chain performance (Fig. 2a). In fact, moving from level 1 to level 4, the performance rises in all the three supply chain structures.

Notice that as the complexity of the SC structure increases, i.e., moving from serial to modular to complex, the performance decreases, regardless of the cognitive capacity level. This trend is expected, because it is known that as K increases, the performance of the system decreases, because the landscape becomes more rugged and multipeaked and consequently the adaptation becomes more complex (Kauffman 1993). This result confirms the validity of the proposed model.

We further quantified the effects of the cognitive capacity level by comparing the results achieved in level 4 against those in level 1. In the case of a serial supply chain, the difference in performance between level 4 and level 1 is about 4.5 %, while in that of a modular supply chain, the difference in performance rises to 7.1 %. The highest difference is achieved in a complex supply chain (about 9.5 %).Thus, the more complex the supply chain structure, the more important the cognitive capacity of the decision maker becomes in terms of the impact on performance.

Results for high resistance to change follow a similar trend. As the cognitive capacity level rises (i.e., moving from level 1 to level 4), performance tends to improve, except for level 3 in the case of serial and modular structures. In such structures, a central planner with a high ability to conceive alternatives but a low cognitive limit (i.e., limited ability to compare a number of alternatives) decreases performance. In particular, performances are very close to those achieved in level 1, i.e., when the central planner has a limited ability to develop alternatives.

Table 4 Performance
difference between No and
High resistance to change

	Serial	Modular	Complex
Mean			
Level 1	0.0068[a]	0.0054[a]	0.0026[b]
Level 2	0.0000[b]	0.0024[b]	0.0004[b]
Level 3	0.0352	0.0394	0.0269
Level 4	0.0099	0.0111	0.0200
Standard deviation			
Level 1	0.0669	0.0752	0.0778
Level 2	0.0638	0.0722	0.0720
Level 3	0.0631	0.0653	0.0651
Level 4	0.0202	0.0344	0.0539

Difference statistically significant with p<0.00001. [a] Significant
with p<0.01. [b] Statistically not significant

Moreover, for simple supply chain structures, where the landscape is not rugged ($K = 2$ and $K = 3$), improving the ability of the decision maker to develop alternatives is not beneficial. Indeed, there is a high probability that developing many alternatives to compare (i.e. 298), results in a move away from the global peak, since the decision maker then chooses one of many different alternatives at random.

On the contrary, high complexity resulting in a multipeak landscape requires that many different alternatives be tried even when the cognitive limit is high, because in such a case the possibility of exploring the landscape (search capability) improves the chances of discovering configurations with a higher performance.

Based on the above, we can affirm that noncomplex supply chain structures, such as serial and modular ones, may improve performance while pursuing integration only by increasing the cognitive limit of the central planner.

Finally, to quantify the effect of the cognitive capacity level, we compared the results achieved at level 1 with those of level 4. The performance difference increased as the supply chain complexity grew. So we confirm that, even when the central planner is resistant to change, it is more important to have a central planner with a high level of cognitive capacity as the complexity of the supply chain structure increases.

We will now discuss the effect of resistance to change of the decision maker on performance. In Table 4, the performance difference between results in cases of no and of high resistance to change is computed for each supply chain structure and for each cognitive capacity level. It can be seen that resistance to change decreases supply chain performance because all differences are positive. This is an expected result because the central planner, even when a configuration with a better performance exists, prefers to maintain the status quo due to his/her risk aversion. Notice that performance differences are however quite low: the highest are achieved in the case of cognitive capacity level 3, whereas the lowest are associated with level 2.

6 Conclusions

This chapter has investigated the benefits of SCI by including behavioral factors influencing decision maker behavior. In fact, the majority of the studies on the topic assume that SCI is pursued by a central planner, which is completely rational and adopts an optimizing behavior. Since the central planner is an individual, such assumptions appear to be too simplistic. In particular, the personal characteristics and motives of the decision maker are important factors to be considered in the analysis because they affect his/her decision-making behavior.

We have focused the attention on two main behavioral factors: (1) cognitive capacity and (2) resistance to change. Managers in fact differ in terms of ability to solve a problem. This depends on their ability to conceive alternatives solutions to test and on their cognitive limit in comparing alternatives and recognizing the best one. Resistance to change is an attitude of individuals who are averse to risk and prefer to not modify/try new solutions for fear of poor outcomes, fear of the unknown, or/and fear of realization of faults.

To pursue our research aim, we developed a simulation analysis in which a central planner characterized by four increasing levels of cognitive capability and two levels of resistance to change is engaged to manage a supply chain in an integrated manner. Three types of supply chain structures characterized by increasing complexity were also considered. Finally supply chain performances were measured and the results compared.

Our simulation analysis has allowed us to show that supply chain performance varies with the cognitive capacity level of the decision maker and his/her resistance to change. In particular, supply chain performance increases as the level of cognitive capacity improves in all the supply chain structures. Moreover, our results have shown that the cognitive capacity of the decision maker becomes more important, in terms of its impact on performance, as the complexity of the supply chain increases. A further result of the study has been that resistance to change decreases supply chain performance even if the effect on performance is quite low.

Thus, further research will be devoted to identify appropriate strategies able to improve supply chain performances mainly in case of low levels of cognitive capacity of the decision maker.

References

R.H. Ballou, S.M. Gilbert, A. Muckherjee, New managerial challenger from supply chain opportunities. Ind. Mark. Manag. **29**, 7–19 (2000)

E. Bendoly, K. Donohue, K. Schultz, Behavior in operations management: Assessing recent findings and revisiting old assumptions. J. Oper. Manag. **24**, 737–752 (2006)

P. Berends, G. Romme, Simulation as a research tool in management studies. Europ. Manag. J. **17**(6), 576–583 (1999)

D.J. Bowersox, D.J. Closs, T.P. Stank, 21st century logistics: Making supply chain integration a reality. Council of logistics management (Oak Brook, 1999)

K.M. Carley, L. Gasser, Computational organizational theory. A modern approach to distributed artificial intelligence, in *Multiagent Systems*, ed. by G. Weiss (The MIT Press, Cambridge, 2000)

T.Y. Choi, K. Dooley, M. Rungtusanatham, Supply networks and complex adaptive systems: Control versus emergence. J. Oper. Manag. **19**, 351–366 (2001)

M. Christopher, *Logistics & Supply Chain Management* (Pitmans, London, 1992)

J.P. Davis, K. Eisenhardt, C.B. Bingham, Developing theory through simulation methods. Acad. Manag. Rev. **32**, 480–499 (2007)

R.D. Dubrin, A.J. Ireland, *Management and Organization, 2nd edn* (South Western Publishing, Cincinnati, 1993)

R. Ernst, B. Kamrad, Evaluation of supply chain structure through modularization and postponement. Eur. J. Oper. Res. **124**, 495–510 (2000)

S. Ethiraj, D. Levinthal, Modularity and innovation in complex systems. Manag. Sci. **50**(2), 159–173 (2004)

M.T. Frohlich, R. Westbrook, Arcs of integration: An international study of supply chain strategies. J. Oper. Manag. **19**(2), 185–200 (2001)

I. Giannoccaro, P. Pontrandolfo, Supply chain coordination by revenue sharing contracts. Int. J. Prod. Econ. **89**(2), 131–139 (2004)

F. Gino, G. Pisano, Toward a theory of behavioural operations. Manuf. Serv. Oper. Manag. **10**, 676–691 (2008)

J.A. Hill, S. Eckerd, D. Wilson, B. Greer, The effect of unethical behaviour on trust in a buyer-supplier relationship: The mediating role of psychological contract violation. J. Oper. Manag. **17**, 281–293 (2009)

S. Kauffman, *The Origins of Order: Self-Organization and Selection in Evolution* (Oxford University, New York, 1993)

J.P. Kotter, Leading change: Why transformation efforts fail. Harv. Busin. Rev. **73**, 59–67 (1995)

D.M. Lambert, M.C. Cooper, J.D. Pagh, Supply chain management: Implementation issues and research opportunities. Int. J. Logist. Manag. **9**(2), 1–19 (1998)

H.L. Lee, V. Padmanabhan, S. Whang, The bullwhip effect in the supply chains. Sloan Manag. Rev. **38**(3), 93–102 (1997)

D.A. Levinthal, Adaptation on rugged landscapes. Manag. Sci. **43**, 934–950 (1997)

F.R. Lin, M.J. Shaw, Re-engineering the order fulfillment process in supply chain network. Int. J. Flex. Manuf. Syst. **10**, 197–229 (1998)

C.H. Loch, Y. Wu, *Behavioral Operations Management* (Now Publishers Inc., Hanover, 2007)

S.P. Mantel, M.V. Tatikonda, Y. Liao, A behavioral study of supply manager decision-making: Factors influencing make versus buy evaluation. J. Oper. Manag. **24**, 822–838 (2006)

E.A. Morash, S.R. Clinton, Supply chain integration: Customer value through collaborative closeness versus operational excellence. J. Market. Theory Pract. **6**(4), 104–120 (1998)

S.D. Pathak, J.M. Day, A. Nair, W.J. Sawaya, M.M. Kristal, Complexity and adaptivity in supply networks: Building supply network theory using a complex adaptive systems perspective. Decis. Sci. **38**, 547–580 (2007)

J.W. Payne, J.R. Bettman, E.J. Johnson, The Adaptive Decision Maker. (Cambridge University Press, New York, 1993)

J.W. Rivkin, Imitation of complex strategies. Manag. Sci. **46**, 824–844 (2000)

J.W. Rivkin, Reproducing knowledge: Replication without imitation at moderate complexity. Organ. Sci. **12**, 274–293 (2001)

J.W. Rivkin, N. Siggelkow, Balancing search and stability: Interdependencies among elements of organizational design. Manag. Sci. **49**, 290–311 (2003)

E. Siemsen, That thing called Be-Op's. POMS-Chronicle **16**, 12–13 (2009)

N.J. Siggelkow, J.W. Rivkin, Speed and search: Design organizations for turbulence and complexity. Organ. Sci. **16**, 101–122 (2005)

D. Simchi-Levi, P. Kaminsky, E. Simchi-Levi, *Designing and managing the supply chain: Concepts* (Strategies and Case Studies McGraw-Hill International, Singapore, 2000)

H.A. Simon, Rational decision making in business organizations. Am. Econ. Rev. **69**(4), 493–513 (1979)

G.C. Stevens, Integrating the Supply Chain. Int. J. Phys. Distrib. & Logist. Manag. **19**(8), 3–8 (1989)

G.N. Stock, N.P. Greis, J.D. Kasarda, Enterprise logistics and supply chain structure: The role of fit. J. Oper. Manag. **18**, 531–547 (2000)

T.J. Sturgeon, Modular production networks: A new American model of industrial organization. Ind. Corp. Change **11**(3), 451–496 (2002)

A. Surana, S. Kumara, M. Greaves, U.N. Raghavan, Supply-chain networks: A complex adaptive systems perspective. Int. J. Prod. Res. **20**, 4235–4265 (2005)

K.T. Ulrich, The role of product architecture in the manufacturing firm. Res. Policy **24**, 419–440 (1995)

S.K. Vickery, J. Jayaram, C. Droge, R. Calantone, The effects of an integrative supply chain strategy on customer service and financial performance: An analysis of direct versus indirect relationships. J. Oper. Manag. **21**, 523–539 (2003)

K.K. Whathne, J.B. Heide, Opportunism in interfirm relationships: Forms, outcomes, and solutions. J. Market. **64**, 36–51 (2000)

X. Zhao, B. Huo, B. Flynn, J.H.Y. Yeung, The impact of power and relationship commitment on the integration between manufacturers and customers in a supply chain. J. Oper. Manag. **26**, 368–388 (2008)

Chapter 6
Cognitive Biases, Heuristics, and Overdesign: An Investigation on the Unconscious Mistakes of Industrial Designers and on Their Effects on Product Offering

Valeria Belvedere, Alberto Grando and Boaz Ronen

Abstract This chapter reports the preliminary findings of an empirical study aimed at understanding whether and to what extent cognitive biases determine overdesign. Overdesign occurs when designers develop product that exceed customers' needs. This phenomenon—which results in higher costs and in some cases also in lower revenues—can be determined by some behavioral problems, as the willingness to develop the "best possible product", regardless of customers' needs. Thus, building on previous studies on cognitive biases, we have conducted a survey among industrial designers, in order to check whether overdesign is driven by cognitive biases. The preliminary evidence shows that this assumption is confirmed. However, the direction of the relationship is negative. This means that the higher the magnitude of the bias, the lower the overdesign. Thus we claim that, in the sample analyzed in this study, we are not in presence of "cognitive biases", but of "heuristics" that can mitigate overdesign. We conclude that designers' experience can be the condition that must occur in order to have a bias turned into a heuristic.

V. Belvedere (✉)
Bocconi University and SDA Bocconi School of Management,
via Roentgen 1 20136 Milan, Italy
e-mail: valeria.belvedere@sdabocconi.it

A. Grando
Bocconi University and SDA Bocconi School of Management,
via Bocconi 8 20136 Milan, Italy
e-mail: alberto.grando@sdabocconi.it

B. Ronen
Leon Recanati Graduate School of Business Administration,
Tel Aviv University, 69978 Tel Aviv, Israel
e-mail: boazr@post.tau.ac.il

I. Giannoccaro (ed.), *Behavioral Issues in Operations Management*,
DOI: 10.1007/978-1-4471-4878-4_6, © Springer-Verlag London 2013

1 Introduction

This chapter presents the preliminary findings of an empirical study aimed at understanding whether and to what extent cognitive biases determine overdesign. Overdesign occurs when designers develop new products whose features exceed the requirements of the customers or of the market. Previous studies on this topic claim that this attitude can be due to organizational problems (i.e., poorly designed performance measurement systems, pricing policies, budgeting rules) and to behavioral ones. While the impact of the former has been already addressed, the latter is still to be analyzed.

Recently, a new stream of research has been started that concerns Behavioral Operations, which aims at incorporating behavioral and cognitive factors in operations management studies. Thus we build on the existing literature on cognitive biases and test the hypothesis according to which cognitive biases are a relevant driver of overdesign.

In the remainder of this paper, a brief literature review is reported. Then the research methodology is explained and the empirical findings and conclusions are drawn.

2 Literature Background

2.1 The Dimensions of Overdesign

Overdesign has been defined as "designing and developing products or services beyond what is required by the specifications and/or the requirements of the customer or the market" (Ronen and Pass 2008; Coman and Ronen 2009).

Building on this definition, two main dimensions of overdesign can be identified. The former has to do with the problem of excessive product variety; the latter refers to the misalignment between the actual performance of the product and the one that customers could be willing to pay for.

According to Ulrich (2010), product variety depends on the combination of three typologies of attributes: fit, taste, and quality. Fit attributes "…are those for which the user's preference exhibits a single strong peak for a single value of the attribute, with satisfaction falling off substantially as the artifact diverges from this value" (Ulrich 2010, p. 115). An example could be the size of a garment. Taste attributes show a multimodal customers' preference function in that the user could have a remarkable preference for a given value of the attribute but at the same time he/she could also praise some alternatives. It is the case of colors for a given garment. Finally, quality attributes are those for which customers would prefer the highest (or lowest) value if this would not have an impact on price. This can be the case of durability for a garment, where customers would theoretically maximize the number of washing cycles if price would not change as a function of this attribute.

If we describe variety in terms of attributes, two measures can be obtained (Ulrich 2010), which refer to the total number of stock-keeping units (SKU) offered by the company and to the number of attributes that each SKU is endowed with. Thus, although numerically two companies can have the same number of SKUs, the complexity involved by their product range can be different if a company has endowed its products with a higher number of attributes. This has great relevance especially in cases where the bill of materials is wide and encompasses several components (Ramdas et al. 2003; Randall and Ulrich 2001).

Thus, building on Ulrich's taxonomy, we can claim that overdesign can be observed from the following perspectives:

- number of SKUs;
- number of attributes per SKU.

It must also be considered that, especially for quality attributes, the management has to choose the level of intensity of the attribute (i.e., if product durability is concerned, the company might be willing to set it at the maximum possible level) and the degree of tolerance around it. When taking these decisions, the company can exceed what the target client wants to receive. This is likely to happen also for fit and taste attributes, where the risk is that the company offers a too wide range of alternatives (e.g., too many colors). Furthermore, designers often set too tight tolerance intervals, which are not consistent with the natural tolerance of the manufacturing process. This leads to treat as a scrap a final product whose actual performance can comply with customers' expectations. Thus, we can claim that overdesign can be also assessed in terms of:

- intensity of the average performance (or number of alternatives) of an attribute.

While the above-mentioned concepts refer to functional overdesign, another dimension of this phenomenon exists that concerns esthetics.

Gaining an insight into the role of the aesthetic dimension in product design is relevant, due to the increasing importance of "designers"—and not just "engineers"—in NPD. In fact, as documented by Perks et al. (2005), in the past years the degree of involvement of designers in this process has remarkably increased and in some cases they act as process leaders. While this evolution can bring a high level of innovation in the product range, especially if a differentiation strategy is pursued, it can be also a threat if the market knowledge of the designer is low. Furthermore, when designers are poorly aware of manufacturing constraints and, namely, of the tolerances of the process, they can take value-destroying decisions (Di Stefano 2006). Thus, on the basis of this evidence, Perks et al. (2005) call for an accurate training of the designers, aimed at endowing them with a wider set of skills and competencies, and also for a recruitment policy focused on the selection of designers with a long experience in the industry.

Although the esthetic dimension (and, generally speaking, the design saliency) is essential in many cases, namely for design-intensive products, few studies have provided a comprehensive framework suitable for understanding how it must be considered within the whole set of "tangible" and/or "measurable" attributes that describe a product. Previous contributions have brought evidence of the impact not only of technical newness but also of the esthetic one on the economic performance of the firm (Talke el al. 2009; Hertenstein et al. 2005). However, it is still hard to perform an analysis on the alignment between the esthetic content of a new product and the one that the customer is willing to pay for.

A major contribution in this regard is the one of Bloch (1995), which is the first notable attempt to observe product form and its effect on customers' response. According to Bloch, "product's form represents a number of elements chosen and blended into a whole by the design team to achieve a particular sensory effect". Although Bloch provides and detailed description of the elements in his framework, "product form" itself remains a kind of "black box" that evokes both esthetics and functionalities, which however are not precisely defined. Noble and Kumar (2010) have tried to expand Bloch's concept of "product form", providing a new reference model. Its most notable feature concerns the fact that, according to these authors, both customers and designers share a value-based view of the product; namely they distinguish among rational, kinesthetic, and emotional value. However, as highlighted by Noble and Kumar, customers and designers can have different perceptions of such value. This can be a source of overdesign.

The evidence brought by Noble and Kumar is confirmed by recent studies in the stream of research concerning the effectiveness of product positioning decisions. Indeed, a driver of this phenomenon could be the way in which customers perceive some specific attributes of the products. However, it has been demonstrated that the way in which customers "see" and praise the product is different from the schematic way commonly used by the company. In this regard, Fuchs and Diamantopoulos (2012) have recently argued that it is almost impossible to predict the positioning success of a new product on the basis of customers' assessment of each distinctive feature of the product. In fact, they claim that most products are endowed with several complex attributes, which cannot be assessed by the customers since they do not have good enough technical competencies. Furthermore, products often have some intangible features (as image and brand identity) that cannot be easily assessed by customers, who, on the contrary, evaluate the product as a whole. The rationale of this paper is actually confirmed by other studies, which demonstrate that, for example, customers praise well-designed objects (Gabrielsen et al. 2010; Kristensen et al. 2012), as well as the presence of visual art in a product (Patrick and Hagtvedt 2011; Hagtvedt and Patrick 2008).

Although the extant literature on product form does not provide specific tools or metrics suitable for measuring this kind of overdesign, it clearly highlights that overdesign is a major issue and that it concerns not only the esthetic dimension of the product but, generally speaking, all its attributes. Furthermore, even though this stream of research does not provide any definition that can be operationalized into an assessment tool, it is possible to build on it and claim that overdesign exists in all cases where:

- the features of the product exceed customers' requirement (or perceptions).

2.2 Behavioral Problems as a Source of Overdesign

Given that overdesign can affect a number of products/industries, it is worthwhile understanding its causes and its consequences. The designers' attitude toward overdesign has been observed and documented especially in technology-based industries, as electronics and IT applications, where rather often companies launch new items that exceed customers' requirements, thus leading to some unexpected and unfavorable outcomes. First of all, firms that experience overdesign suffer from long times to market, which often result in delays in the product launch. In time-based industries, as consumer electronics, this can determine a poor economic and market performance of the new product, due to a problem of rapid technical obsolescence. Second, when overdesign takes place, products tend to be too complex and customers are not able to properly use them. This turns into poor customer satisfaction and, in some cases, also in damages and in subsequent returns. Furthermore, when designers add to the product excessive features, so as to make it suitable for any potential customer (and not just for the target one), the selling price is generally higher, with a negative effect on the market share that the product can reach in its target segment. All of these unfavorable effects of overdesign can destroy a company's value, thus it is a major issue to understand why this phenomenon takes place and how it can be reduced.

According to previous studies (Coman and Ronen 2009; Ronen and Pass 2008), overdesign has several sources, which can be summarized as follows:

- *behavioral problems*. As maintained by Simon (1957), managers tend to have an *Optimizer Approach* to their work activity. Indeed, in many different fields people struggle to achieve the best possible solution, regardless of the negative effect that this can have on the amount of resources necessary to reach this aim. In R&D projects this approach is rather common. In fact, developing the "best possible" product can theoretically bring about some potential benefits, as the possibility to reach a larger part of the market or to anticipate some future evolutions in customers' requirements. However, this approach has proved to be ineffective in several cases (Coman and Ronen 2009). Previous studies claim that it is rooted in a lack of knowledge of the market and of the manufacturing processes, common to R&D people, who are willing to enrich the product with a number of features that the client is not interested in or that can be hardly obtained with the available equipment and machinery (Coman and Ronen 2009). The Optimizer Approach is also due to a problem of culture, since designers and engineers measure their own professional success on the basis of the technological performance of their products rather than on the basis of the value created by them;
- *organizational problems*. Some organizational mechanisms, as performance measurement, pricing policies, and budgeting rules, are relevant drivers of overdesign (Ronen and Pass 2008). As extensively proved in the literature on these issues, people's behavior is strongly influenced by the way in which they are assessed. Thus, if the designer's professional performance is measured according to the number of new products conceived, he/she will tend to work on

as many projects as possible, thus boosting overdesign. Also, the pricing policy can be a driver of this phenomenon. If the selling price of a product is defined through a "cost plus" approach rather than through a value analysis, designers will be less concerned with the total cost incurred by the company to develop and launch the new product. Finally, some budgeting procedures can lead toward overdesign. In companies where R&D financial resources can be allocated only to customers projects, designers and engineers willing to work on new technologies have to embed them in the new products in order to obtain a budget.

While the second source of overdesign has been addressed by several studies in the fields of performance measurement, marketing, and accounting, the first one (i.e., behavioral problems) has recently become popular among management scholars. Building on the seminal works of Kahneman and Tversky and on the Prospect Theory conceived by these two authors (Kahneman and Tversky 1979), new streams of research have been started, aimed at incorporating behavioral and cognitive factors in management studies. Recently, a behavioral perspective has been adopted also in the field of operations management, where the opportunity to analyze the cognitive issues peculiar to product development and project management has been highlighted (Gino and Pisano 2008). In fact, as claimed by Kahneman and Tversky, human beings suffer from a number of cognitive biases and frequently adopt heuristics (availability, representation, anchor-and-adjustment) that often lead them to take irrational decisions (Kahneman and Tversky 1974, 1979). Also, the way in which people *frame* their decisions can lead to contradictive and counterintuitive outcomes (Kahneman and Tversky 1981). Building on this approach, it can be argued that some of the behavioral problems that lead toward overdesign can be analyzed moving from the contributions of Kahneman and Tversky. Although this can be an innovative and fruitful approach, most studies that adopt this perspective are based on experiments that are not carried out specifically in R&D teams. Thus, it is necessary to understand how these heuristics and cognitive biases can be defined and measured in such an environment before analyzing the problem of overdesign from this perspective.

In this regard, two recent contributions seem to be the most relevant, i.e. Lovallo and Sibony (2010) and Kahneman et al. (2011). The former contribution builds on the idea that executives can be aware of the cognitive biases that affect their choices, however, they might not know how to "debias" decision making processes. On the basis of an extensive survey, Lovallo and Sibony (2010) identify five key typologies of cognitive biases peculiar to executives and propose precise definitions for each of them:

- *action-oriented biases*. They concern the excessive optimism that often drives decision making and that is often accompanied by the tendency of neglecting competitive responses;
- *interest biases*. They take place in presence of conflicting incentives, namely at corporate and functional levels;

- *pattern-recognition biases*. They take place when decision making is heavily based on past experiences;
- *stability biases*. They concern the tendency toward inertia in case of uncertainty;
- *social biases*. They arise when people have a preference for harmony within the group rather than for discussion of counterarguments.

The latter contribution (Kahneman et al. 2011) has further developed the idea of Lovallo and Sibony (2010) and has operationalized it into a self-assessment of 12 questions to let companies (or groups) understand whether they suffer from some specific cognitive biases.

Although these two contributions are not specifically tailored on product development and project management, nevertheless they often discuss cases concerning R&D environments. Thus, it can be argued that the cognitive biases as defined by Lovallo and Sibony (2010) and by Kahneman et al. (2011) can be used to study the impact of behavioral problems on R&D activities and, namely, on the phenomenon of overdesign.

This kind of analysis can be useful, since there is not any study that has tried to check and to quantify the extent to which cognitive biases determine overdesign. This analysis can be fruitful because, as demonstrated by Lovallo and Sibony (2010), companies, which are aware of their biases and are able to "debias" their decision making processes, are likely to reach a higher level of effectiveness.

3 Research Question and Methodology

On the basis of the literature analysis, we wanted to carry out an empirical investigation aimed at understanding *whether and to what extent cognitive biases determine overdesign*.

Building on previous contributions, we developed the reference framework described in Fig. 1.

According to this framework, the phenomenon of overdesign can be explained by a bundle of cognitive biases. Namely, we described such biases using the five typologies identified by Lovallo and Sibony (2010). Furthermore, we assumed that the direction of this influence should be positive, so that the higher the magnitude of the bias, the higher the overdesign. This assumption is consistent with previous studies on cognitive biases, which consider them as negative factors that can lead human beings toward irrational decisions. Lovallo and Sibony (2010) and also Kahneman et al. (2011) present biases as phenomena that should be removed in order to improve the effectiveness of the decision-making processes. In fact, Lovallo and Sibony (2010) clearly point out the necessity for companies to "debias" these processes so as to foster their overall effectiveness.

Given the nature of the research question, we decided to adopt a quantitative methodology and, namely, to carry out a survey.

Fig. 1 The reference
framework

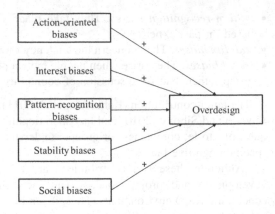

To operationalize our framework, we developed a questionnaire made of 24
statements reported in Table 1. For all of them, assessments on a Likert scale from
1 ("totally disagree") to 7 ("totally agree") have been requested. For item 10, the
scale has been later reversed, so as to make it consistent with the construct to be
measured.

The statements concerning cognitive biases (from 8 to 24 in Table 1) have been
based on Lovallo and Sibony (2010) and Kahneman et al. (2011). For all of them,
we have built on the original statements proposed by these authors, just adapting
their contents to a typical NPD environment.

For the "Overdesign" construct, we have built on a wider range of extant
contributions. Indeed, given that overdesign has not been operationalized into
specific dimensions yet, we have developed seven statements (namely, items 1–7
in Table 1) that capture the perspectives from which this construct can be
observed. In particular, questionnaire items 1 and 2 build on Ronen and Pass
(2008) and on Coman and Ronen (2009), according to whom overdesign is a
pathology that results in a long time to market and in frequent delays in the
innovation projects. Items 3 and 4 are also based on Ronen and Pass (2008) and on
Coman and Ronen (2009) and on the literature about "product form", which
highlights how designers tend to enrich their products with functional and/or
esthetic features to pursue a differentiation strategy or just for a lack of market
knowledge (Perks et al. 2005).

Items 5, 6, and 7 build on Ulrich (2010) and aim at assessing overdesign from
the perspective of product attributes. Namely, item 5 explicitly refers to the
attributes ("features") of the product and to the possibility that they can exceed
customers' needs. Items 6 and 7 concern product range (which results from all the
possible combinations of the attributes) and point out the problem of its constant
renewal (item 6) and widening (item 7).

The questionnaire has been sent by e-mail to the members of the Italian
Association of Industrial Engineers (AIPI). It totally counts 264 members, but only
187 have provided the Association with their e-mail address. The questionnaire
collection was started in November 2011 and, at the end of December 2011, 47

Table 1 Constructs and questionnaire items

Constructs	Statements
Overdesign	(1) Our projects are never delivered on time
	(2) The milestones of our projects are never met
	(3) Our products are ahead of our competitors
	(4) Most of our effort in developing new products is driven by the needs of potential new customers
	(5) The features of our products always exceed customers' requirements
	(6) Our product range is constantly enriched and renewed. We always add new products to our offering
	(7) The number of products is constantly growing
Action-oriented biases	(8) The members of the team are, in general, optimistic about the outcome of planned actions and of the overall project
	(9) The members of the team are, in general, optimistic about the skills of the team itself compared to those of the competitors. They think that these skills can lead to positive outcomes of planned actions and of the overall project.
	(10) In assessing the success of the new product, all necessary information is gathered, including those concerning how competitors will react
Interest biases	(11) When assessing and conducting a project for a new product, team members often pursue individual (or team) goals rather than corporate goals
	(12) When working on a project, team members seem to be emotionally attached to the new product they are developing
	(13) When assessing and conducting a project for a new product, team members are unclear about corporate goals, their hierarchy and possible trade-offs
Pattern-recognition biases	(14) When assessing a new product or project, team members give more relevance to evidence and information that support the project rather to those that can lead to a negative judgment
	(15) When assessing a new product or project, team members recall recent or memorable examples
	(16) When assessing a new product or project, team members recall examples or stories that are frequently told in their company
	(17) When assessing a new product or project, team members give more relevance to the track record of the person presenting the project than to the evidence that supports it
Stability biases	(18) When assessing a project or a product (e.g. margins generated by the product, market share, time necessary to complete the project, number of designers required etc.), the team (or the project leader) moves from a reference value defined on the basis of available information (historical data, competitors etc.) and then adjusts it
	(19) In the overall assessment of a project or product, team members are more cautious when facing a risk of lower future profits
	(20) In the overall assessment of a project or product, team members often undertake actions that increase development costs, if they are suitable for improving the performance of the product
	(21) When deciding whether to complete a project or not, the team members pay much attention to the sunk costs
	(22) In managing projects, the team tends to replicate practices and decision-patterns already experimented in the past
Social biases	(23) In managing the development team, a major importance is given to the consensus of the team members in the overall assessment of the project or product
	(24) Team members tend to agree on the viewpoint of the team leader

questionnaires were collected. Then, a new mailing has been sent to other associations of designers in early 2012 and the questionnaire collection process is still
in progress. This paper reports the preliminary findings of this research project,
based on the data collected in 2011 from AIPI members.

4 Empirical Evidence

To test the validity of the reference framework reported in Fig. 1, a factor analysis
and then a regression analysis have been conducted. The former step was undertaken in order to check whether the factors coming out from the analysis are
consistent with the reference model. The latter aimed at testing the hypothesis that
cognitive biases are relevant predictors of overdesign.

As far as the factor analysis is concerned, we retained only the factors with an
Eigenvalue higher than 1 (Hair et al. 2006). Items have been retained if their factor
loading was higher than 0,5 only for a single factor and lower than this threshold
for all the others (Hu and Bentler 1999; Stevens 1986).Table 2 reports the questionnaire items and the corresponding factor loadings.

As it can be seen, this analysis highlights the existence of a factor that seems to
describe the "overdesign" construct (Factor 1). Moving to the other factors, it can
be seen that group cognitive biases in a way that is slightly different from the
typologies defined by Lovallo and Sibony (2010). The reliability of the scale
adopted to study these phenomena has been tested computing the Cronbach's
alpha for each factor. These values, being all higher than 0,7, bear witness to the
reliability of this measurement instrument (Nunnally 1978).

Table 2 Rotated factor pattern

	Factor 1	Factor 2	Factor 3	Factor 4	Factor 5	Factor 6	Factor 7
IT1	0,811						
IT2	0,812						
IT6	−0,680						
IT7	0,744						
IT8		0,766					
IT9		0,699					
IT20		0,699					
IT15			0,781				
IT16			0,784				
IT22			0,676				
IT14				0,716			
IT17				0,702			
IT24				0,672			
IT10					0,730		
IT23					0,749		
IT18						0,837	
IT21							0,832

Factor 1 encompasses the following questionnaire items:

IT1. Our projects are never delivered on time

IT2. The milestones of our projects are never met

IT6. Our product range is constantly enriched and renewed. We always add new products to our offering

IT7. The number of products is constantly growing

Such a factor clearly refers to overdesign. In fact, two items (1 and 2) concern the problem of long time to market and frequent delays that, according to Ronen and Pass (2008) and to Coman and Ronen (2009), are a typical negative consequence of overdesign. Items 6 and 7 describe overdesign from the perspective of product range (Ulrich 2010) and, namely, point out the problem of its frequent renewal and widening.

In order to check whether and to what extent Factor 1 (i.e., Overdesign) depends on cognitive biases, a regression analysis was conducted using the six remaining factors as determinants of "Overdesign". The outcome is reported in Fig. 2.

It can be noted that Factors from 2 to 7 (which express different typologies of cognitive biases) explain 48 % of the variance of Overdesign, with a very good level of statistical significance (p value $< 0,001$). Focusing only on factors that are statistically significant drivers of overdesign, some interesting conclusions can be drawn on their meaning and on the kind of effect that they produce.

Factor 3 seems to express a kind of "backward looking approach", since it is described by the following questionnaire items:

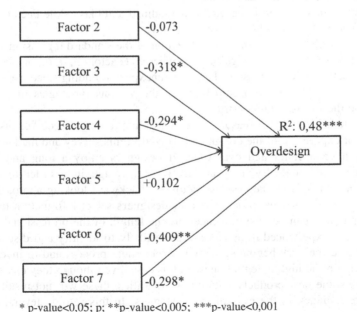

* p-value<0,05; p; **p-value<0,005; ***p-value<0,001

Fig. 2 Evidence from the regression analysis

IT15. *When assessing a new product or project, team members recall recent or memorable examples*

IT16. *When assessing a new product or project, team members recall examples or stories that are frequently told in their company*

IT22. *In managing projects, the team tends to replicate practices and decision-patterns already experimented in the past*

Factor 4 seems to refer to the way in which designers select and interpret information suitable for assessing new products which they are working on. The statements that describe this construct are as follows:

IT14. *When assessing a new product or project, team members give more relevance to evidence and information that support the project rather to those that can lead to a negative judgment*

IT17. *When assessing a new product or project, team members give more relevance to the track record of the person presenting the project than to the evidence that supports it*

IT24. *Team members tend to agree on the viewpoint of the team leader*

Factors 6 and 7 are described by a single statement each:

IT18. *When assessing a project or a product (e.g., margins generated by the product, market share, time necessary to complete the project, number of designers required etc.), the team (or the project leader) moves from a reference value defined on the basis of available information (historical data, competitors etc.) and then adjusts it*

IT21. *When deciding whether to complete a project or not, the team members pay much attention to the sunk costs*

Factor 6 seems to refer to the anchor-and-adjustment bias, while Factor 7 seems to be related to the so-called sunk-cost fallacy.

What is worthwhile noticing is the direction of the standard regression weights. As a matter of fact, the four significant predictors (Factors 3, 4, 6, and 7) show a negative coefficient of regression. This is an unexpected finding since we expected to observe a positive relationship between the various typologies of cognitive biases and the Overdesign construct.

For Factors 3, 4, and 6 a reason for such negative regression coefficients can be the level of experience of the designers involved in this survey and the know-how that they have accumulated over time. If designers enjoy a long and fruitful experience in the industry, they are likely to leverage it in order to prevent overdesign. This can be the case of Factor 3 ("backward looking approach").

Factor 4, which seems to describe how designers select information to assess the new product, can be also influenced by the length of the professional experience. In fact, experienced designers are more likely to identify and discard poor projects since the very beginning of the development process and to invest their efforts only on the high potential ones. In this case, even though they can seem to "sponsor" some new products and/or functionalities, these will not result in any kind of redundancy in the eyes of the customers. In this setting, leveraging not only the personal experience and intuition, but also the ones in the rest of the group and of the team leader can further boost the ability of an NPD team to rapidly identify projects with a high potential. By contrast, a young designer can have the same attitude toward the selection of information, but without a long enough length of service and an experienced group to work with, he/she is likely to keep working on value destroying projects because he/she has not been able to recognize them.

Factor 6 expresses the phenomenon called "anchor-and-adjustment". If designers are able to exploit their experience so as to set reasonable "anchors" for their analyses and decisions, they are likely to identify and reject value destroying projects.

These interpretations for Factors 3, 4, and 6 are backed by the mean age of the respondents (equal to 49 years), the mean length of stay in the current company (15 years) and in the current industry (25 years).

Finally, the negative regression coefficient of Factor 7 can depend on the fact that if sampled designers have a strong commitment to "value creation" and select their projects accordingly, they are likely to rapidly identify bad projects and reject them. If this holds, only a few, good projects are launched and, for these, a strong commitment to value creation and to cost saving is observed.

These intuitions, based on the preliminary findings of this study, show that, rather than being in presence of "biases" (which are supposed to have a negative effect on decision-making processes), we observe the existence of effective heuristics. This is consistent with the recent contribution of Gigerenzer, according to whom human beings tend to adapt themselves to the environment in which they live and to develop "rules of thumb" that, in specific conditions and situations, prove to be effective (Goldstein and Gigerenzer 2009; Gigerenzer and Selten 2001). This concept has been called "ecological rationality" (Gigerenzer 2000). Consistently with this assumption, Gigerenzer has demonstrated the existence of several heuristics, which are frequently adopted in a number of situations and of professional environments, as justice, medicine, and management. One of the most relevant heuristic is the "recognition" one, which refers to the ability of rapidly reach ("recognize") the correct solution to a problem (Goldstein and Gigerenzer 2002). A distinctive feature of such heuristic is that people who adopt them generally use a few pieces of information and apparently decide on the basis of an "instinct" (Goldstein and Gigerenzer 2009).

In fact, this new perspective on biases and heuristics has been adopted also by Kahneman, who, in a recent publication (Kahneman 2011) explicitly claims that intuition and rationality coexist in human beings and that good decisions should involve both of them. Kahneman calls these two approaches to decision-making "System 1" and "System 2". The former is emotional, intuitive and fast. By contrast, the latter is more logical and slower. Kahneman argues that in some cases System 1 (i.e., intuitions and emotions) can lead to effective and fast decisions. However, in order to prevent some biases and distortions that System 1 can determine, its decisions must be somewhat validated by System 2 (i.e., rational analysis).

This can be the case of our sample, in which it is likely that respondents, leveraging their long working experience, have developed the ability to recognize products with a good market potential. Furthermore, it can be assumed that this ability is not a peculiarity of the single designer but of the group as a whole. In this concern, "storing" knowledge in the form of memorable examples and stories and then "retrieving" them when an analogy is found can be an effective way to leverage designers' intuitions. In fact, this organizational mechanism implies that, while the analogy can be intuitively identified by an individual, the final assessment on the product is based on a discussion with the team members. This can be actually a rational "gate" that leads toward sound decisions and that works properly if all team members share a deep knowledge of the product and of the process.

5 Conclusions

This paper presents the preliminary findings of a survey aimed at understanding *whether and to what extent cognitive biases determine overdesign.* The evidence reported is based on 47 questionnaires collected among the members of the Italian Association of Industrial Engineers. However, since the collection is still in progress, the final outcome of this research project could be slightly different from that reported in this paper. Based on these preliminary findings, we can confirm that cognitive biases are a relevant driver of overdesign, but two interesting outcomes arise from this study. First of all, we initially moved from five typologies of biases, based on previous studies on this topic. Our study reveals that at least in R&D environments it is still necessary to clearly define the nature of such biases. Second, it comes out that some of these biases are negatively correlated with the phenomenon of overdesign, in that when they occur, overdesign decreases. This evidence needs to be confirmed analyzing other datasets. However, if it is confirmed, it could be worthwhile studying why this counterintuitive relationship is observed. An explanation builds on the level of experience and of know-how of the designers, which can bring about some effective heuristics (and not "biases") that can help in mitigating overdesign. If this evidence is confirmed in future studies, then it could be worthwhile understanding the conditions under which cognitive biases can be turned into effective heuristics.

References

P.H. Bloch, Seeking the ideal form: product design and consumer response. J. Mark. **59**, 16–19 (1995)

A. Coman, B. Ronen, Icarus' predicament: Managing the pathologies of overspecification and overdesign. Int. J. Project Manag. **28**(3), 237–244 (2009)

P.P. Di Stefano, Tolerances analysis and cost evaluation for product life cycle. Int. J. Prod. Res. **44**(10), 1943–1961 (2006)

C. Fuchs, A. Diamantopoulos, Customer-perceived positioning effectiveness: conceptualization, operationalization, and implications for new product managers. J. Prod. Innov. Manag. **29**(2), 229–244 (2012)

G. Gabrielsen, T. Kristensen, J. Zaichkowsky, Whose design is it anyway? Int. J. Market Res. **52**(1), 89–110 (2010)

G. Gigerenzer, *Adaptive Thinking. Rationality in the Real World* (Oxford University Press, New York, 2000)

G. Gigerenzer, R. Selten (eds.), *Bounded Rationality: The Adaptive Toolbox* (MIT Press, Cambridge, 2001)

F. Gino, G. Pisano, Toward a theory of behavioral operations. Manuf. Service Oper. Manag. **10**(4), 676–691 (2008)

D.G. Goldstein, G. Gigerenzer, Fast and frugal forecasting. Int. J. Forecast. **25**(4), 760–772 (2009)

D.G. Goldstein, G. Gigerenzer, Models of ecological rationality: The recognition heuristic. Psychol. Rev. **109**(1), 75–90 (2002)

H. Hagtvedt, V. Patrick, Art infusion: The influence of visual art on the perception and evaluation of consumer products. J. Mark. Res. **45**(3), 379–389 (2008)

J.F. Hair, W.C. Black, B.J. Babin et al., Multivariate Data Analysis, 6th edn. (Prentice Hall, Upper Saddle River, 2006)

J.H. Hertenstein, M.B. Platt, R.W. Veryzer, The impact of industrial design effectiveness on corporate financial performance. J. Prod. Innov. Manag. **22**(1), 3–21 (2005)

L. Hu, P.M. Bentler, Cutoff criteria for fit indexes in covariance structure analysis: Coventional criteria versus new alternatives. Struct. Equ. Model. **6**(1), 1–55 (1999)

D. Kahneman, *Thinking, Fast and Slow* (Farrar Straus Giraux, New York, 2011)

D. Kahneman, D. Lovallo, O. Sibony, Before you make that big decision. Harv. Bus. Rev. **89**(6), 50–60 (2011)

D. Kahneman, A. Tversky, Judgment under uncertainty: Heuristics and biases. Science **185**, 1124–1131 (1974)

D. Kahneman, A. Tversky, Prospect theory: An analysis of decisions under risk. Econometrica **47**, 263–291 (1979)

D. Kahneman, A. Tversky, The framing of decisions and the psychology of choice. Science **211**, 453–458 (1981)

T. Kristensen, G. Gabrielsen, J. Zaichkowsky, How valuable is a well-crafted design and name brand? Recognition and willingness to pay. J. Consum. Behav. **11**(1), 44–55 (2012)

D. Lovallo, O. Sibony, The case for behavioral strategy. McKinsey Q. **2**, 30–43 (2010)

C.H. Noble, M. Kumar, Exploring the appeal of product design: A grounded, value-based model of key design elements and relationships. J. Prod. Innov. Manag. **27**, 640–657 (2010)

J.C. Nunnally, *Psychometric Theory*, 2nd edn. (McGraw-Hill, New York, 1978)

V. Patrick, H. Hagtvedt, Aesthetic incongruity resolution. J. Mark. Res. **48**(2), 393–402 (2011)

H. Perks, R. Cooper, C. Jones, Characterizing the role of design in new product development: An empirically derived taxonomy. J. Prod. Innov. Manag. **22**(2), 111–127 (2005)

K. Ramdas, M. Fisher, K. Ulrich, Managing variety for assembled products: Modeling component systems sharing. Manuf. Service Oper. Manag. **5**(2), 142 (2003)

T. Randall, K. Ulrich, Product variety, supply chain structure, and firm performance: Analysis of the U.S. bicycle industry. Manag. Sci. **47**(12), 1588–1604 (2001)

B. Ronen, S. Pass, *Focused operations management: Doing more with existing resources* (John Wiley and Sons, New York, 2008)

H.A. Simon, *Models of Man* (John Wiley and Sons, New York, 1957)

J. Stevens, *Applied Multivariate Statistics for the Social Sciences* (Lawrence Erlbaum Associates, Hillsdale, 1986)

K. Talke, S. Salomo, J.E. Wieringa et al., What about design newness? Investigating the relevance of a neglected dimension of product innovativeness. J. Prod. Innov. Manag. **26**(6), 601–615 (2009)

K.T. Ulrich, *Design. Creation of Artifacts in Society* (University of Pennsylvania, Pennsylvania, 2010)

Chapter 7
Incentives in Organizations: Can Economics and Psychology Coexist in Human Resources Management?

Ugo Merlone

Abstract Several disciplines have approached Human Resources Management (HRM) from different perspectives with several authors providing interesting contributions but often ignoring each other. In this chapter, I will illustrate how Economics and Psychology approach HRM in order to point out differences but also to areas of intersection between these two important disciplines. I will mainly concentrate on Economics and Psychology, not for the sake of some alleged priority or superiority of these disciplines, but rather for being more familiar with this literature. In the future research it would be interesting to add perspectives coming from other disciplines. Psychology and Economics, as social sciences, often deal with models of human behavior. It is well known that these two disciplines offer contrasting theories of human behavior on virtually every major point (DeAngelo et al. 2011). One strong point of disagreement is rationality (Smith 1991). We will see that the contrast on rationality is not limited to Economics and Psychology. Even if the differences between these two disciplines are often stark, they both agree on incompleteness of contracts. In this chapter, I will explore how these disciplines approach this gap, what are the main differences, and the possible integrations.

Disciplines are categories that facilitate filing the content of science. They are nothing more than filing categories. Nature is not organized the way our knowledge of it is. Furthermore, the body of scientific knowledge can, and has been, organized in different ways. No one way has ontological priority. Russell L. Ackoff (1973, p. 667).

I am grateful to Laura Basta, Gian-Italo Bischi, Arianna Dal Forno, Paolo Bertoletti, Denise Rousseau, Pietro Terna and Gabriella Valentino for helpful suggestions. Usual caveats apply.

U. Merlone (✉)
Department of Psychology, University of Torino, Via Verdi 10, Torino 10124, Italy
e-mail: ugo.merlone@unito.it

Several disciplines have approached Human Resources Management (HRM) from different perspectives with several authors providing interesting contributions but often ignoring each other. In this chapter I will illustrate how Economics and Psychology approach HRM in order to point to differences but also to areas of intersection between these two important disciplines. I will mainly concentrate on Economics and Psychology, not for the sake of some alleged priority or superiority of these disciplines, but rather for being more familiar with this literature. But of course, it would be interesting to add perspectives coming from other disciplines.

Psychology and Economics, as social sciences, often deal with models of human behavior. It is well known that these two disciplines offer contrasting theories of human behavior on virtually every major point (DeAngelo et al. 2011). One strong point of disagreement is rationality (Smith 1991). We will see that the contrast on rationality is not limited to Economics and Psychology. Even if the differences between these two disciplines are often stark, they both agree on incompleteness of contracts. In this chapter, I will explore how these disciplines approach this gap, what the main differences are, and the possible integrations.

1 Human Resource Management

A first important aspect lies in the definition of HRM. In the late 1980s, Guest (1987) argued that although the term was widely used, its definition was lost. In the Management literature fads are not uncommon, and often new terms rise to popularity (Birnbaum, 2000). For example, Carson et al. (2000) describe fads as "managerial interventions which appear to be innovative, rational, and functional and are aimed at encouraging better organizational performance." According to Boudreau et al. (2003), the HRM approach is derived from disciplines such as Psychology, Sociology, and Inferential Statistics. HRM models describe employment and behavioral processes and their relationships to aspects as rewards, recognition, staffing, sourcing, learning, development, as well as organization structures. HRM focuses on predicting and explaining outcomes such as performance, attraction, retention, loyalty, and citizenship.

According to Legge (2005), although when considering HRM it is possible to find several different models, from the majority of the normative definitions two—not necessarily incompatible—emphases, can be identified. On the one hand HRM should focus on the crucial importance of the close integration of human resource policies and activities with business strategy; in this view human resources (HR) are considered as a factor of production. On the other hand, whereas keeping the importance of integrating HR policies with business objectives, employees are treated as valued assets and as a source of competitive advantage through their commitment, adaptability, and high quality of performance. In this sense humans are not machines and therefore an interdisciplinary examination of people in the workplace is in order. Therefore, several disciplines such as Psychology, Industrial Relations, Industrial Engineering, Sociology, Economics, are called to approach HRM.

Recently, Boudreau et al. (2003) explored the interface between operations management (OM) and human resources by examining how human considerations affect classical OM results and how operational considerations affect classical HRM results. OM deals with the design and management of products, processes, services, and supply chains, and may be defined as the study of decision making in the operations function (Schroeder 1993). The OM approach is grounded on some assumptions which greatly simplify human behavior. They are

(1) "People are not a major factor (Many models look at machines without people, so the human side is omitted entirely).
(2) People are deterministic and predictable. People have perfect availability (no breaks, absenteeism, etc.). Task times are deterministic. Mistakes do not happen, or mistakes occur randomly. Workers are identical (work at the same speed, have the same values, and respond to the same incentives).
(3) Workers are independent (not affected by each other, physically or psychologically).
(4) Workers are "stationary." No learning, tiredness, or other patterns exist. Problem solving is not considered.
(5) Workers are not part of the product or service. Workers support the "product" (e.g., by making it, repairing equipment, etc.) but are not considered explicitly as part of the customer experience. The impact of the system structure on how customers interact with workers is ignored.
(6) Workers are emotionless and unaffected by factors such as pride, loyalty, and embarrassment.
(7) Work is perfectly observable. Measurement error is ignored. No consideration is given to the possibility that observation changes performance (Hawthorne effect)." (Boudreau et al. 2003, p. 183).

Although assumptions (1), (5), and (7) are more related to the technological aspects of the process, it is immediate to see that assumptions (2–4) and (6) are strikingly similar to those used in Economics. In fact, in basic economics, labor is a commodity; the employer buys it at the current market price assuming a definite relation between employees' hours of work and the labor which is provided. This approach is really simplistic and has been criticized; for example, Simon observed that "This way of viewing the employment contract and the management of labor involves a very high order of abstraction-such a high order, in fact, as to leave out of account the most striking empirical facts of the situation as we observe it in the real world. In particular, it abstracts away the most obvious peculiarities of the employment contract, those which distinguish it from other kinds of contracts" (Simon 1951, p. 293).

As employees are not identical machines and the relation between employees' hours of work and labor cannot be realistically considered as fixed, then the problem of incentives arises (Baron and Kreps 1999). For example, when the employer is unable to observe the employee, the latter can provide less work than expected and claim that some contingencies prevented him/her from providing the agreed amount of work. In many cases the employee has different information from the employer, for instance he may be better informed than the employer about production technology. The models considered in the theory of contracts take into account some of the strategic interactions between privately informed agents in well-defined settings.

2 Contract Theory

According to Salanie (1997), contract theory originates in some failures of general equilibrium theory. In fact, whereas general equilibrium theory is one of the most impressive achievements in the history of economic thought (Salanie 1997, p. 1), it appeared that this model was not a fully satisfactory descriptive tool. Among the limitations was the fact that agents were assumed to interact through the price system. This limitation made it difficult to consider models of firms and other economic institutions because, according to Coase (1937), "the distinguishing mark of the firm is the supersection of the price mechanism". Information asymmetry was another limitation for general equilibrium models (Salanie 1997). In fact, although Arrow and Debreau showed how it is possible to extend the general equilibrium theory[1] to cover uncertainty as long as information remains symmetric, often in economic interactions information asymmetries are pervasive. For example, the principal may not be able to observe the action of the agent, employees know more about their cost than employers, or the abilities of a worker may be difficult to observe when the principal is designing an incentive scheme. Some authors provide a classification of asymmetric problems to identify the influence of the nature of the distribution of information about important aspects of the relationship [see for example Macho-Stadler and Pérez-Castrillo (1997) or Laffont and Martimort (2002)]. Economists have created a collection of models to simplify the study of bargaining under asymmetric information by allocating all bargaining power to one of the parties (Salanie 1997). As Baron and Kreps (1999) point out, this collection of models is called *principal-agent model*, *agency theory* and *economic theory of incentives*.

Many kinds of incentive problems can be modeled using a common framework which can adapt to several different situations. Since the focus of this chapter is HRM, we will consider the two parties to be an employer and an employee. Remaining in the focus of HRM, the employer may be a manager hiring a worker or a company owner hiring a manager (Bolton and Dewatripont 2005) but the number of applications is really wide; for instance, agency theory has been used also to address outsourcing relationships (Logan 2000). A simple example of the principal-agent problem is that of an employer who wants the employee to exert as much effort as possible, in order to produce as much output as possible, although

---

[1] For a concise introduction to Arrow-Debrau model of general equilibrium, the reader may refer to Geanokoplos (2004).

the employee rationally wants to make a choice that maximizes his own utility given the effort and incentive scheme. This conflict between the employer and the employee is more evident assuming that the connection between time and effort exerted by the employee and the results in terms of output are not entirely under his control. Assuming that the effort is not directly measurable, while the employee prefers to be paid according to the hours worked, the employer would like to pay according to the employee's output.

From the technical point of view, as Salanie (1997) observes, the principal-agent model is a Stackelberg game in which the one who proposes the contract—the leader—is called the principal and the party who has to accept or reject the contract—the follower—is called the agent.

The basic model of agency relies on three more or less implicit assumptions which, in the case of our example, can be expressed as follows:

1. The employee is averse to the effort. For example, if he is paid on a per-unit-of-time basis, he will choose to exert the lowest level of effort which still allows him to be paid.
2. The employee is risk-averse. That is, if the employee were risk-neutral he would bear all the risks and would completely internalize all the consequences of his choices of effort.
3. The parties cannot contract on the level of effort. In this case, if the employer were risk neutral, she would pay the employee a wage depending simply on the effort, and therefore obtain the efficient outcome.

For a thorough discussion of these assumptions the reader may refer to Baron and Kreps (1999).

Furthermore, as in most Economics literature, strong assumptions on rationality of actors are made; as a fact, Bolton and Dewatripont (2005, p. 5) clearly state "we shall assume that contracting parties are rational individuals who aim to achieve the highest possible payoff". It must be noted that this is the common approach for Economics when analyzing human resources and compensation issues. For example, Milgrom and Roberts (1992, p. 326) state "we assume rational and largely self-interested behavior; we presume that people seek efficient solutions to the problems they face". Whereas, on the one side, this assumption is necessary as a starting point, it is in strong contrast with empirical evidence. The recent interest in Behavioral Economics and the empirical evidence of experimental economics challenge this assumption (for a survey of recent developments the reader may refer to Della Vigna 2009). Yet, as it concerns behavior in organizations, Simon (1997) considers distinct types of rationality and provides some motivational links between the individual and the organization explaining how organizational influences may be effective forces in molding human behavior. Furthermore, his principle of bounded rationality contrasts the above assumptions: "The capacity of the human mind for formulating and solving complex problems is very small compared with the size of the problems whose solution is required for objectively rational behavior in the real world- or even for a reasonable approximation to such objective rationality" (Simon 1957, p. 198).

Several experiments in the Economics and Psychological literature provide evidence that individuals show a concern for the welfare of others (Fehr and Gächter 2000; Charness and Rabin 2002). Finally, there is evidence that individuals may decide to pay some of their own money in order to reduce other's money (Zizzo and Oswald 2001; Divotti and Merlone 2011). Another field evidence shows how individuals use heuristics to solve complex problems (Gabaix et al. 2006) and are affected by emotions in their decisions (Loewenstein and Lerner 2003).

In particular, some field evidence is quite interesting as it concerns workplace relations. Mas (2006) studies the impact of reference points for the New Jersey Police with a relationship between police pay and the share of solved crimes; Krueger and Mas (2004) examine how the quality of tyres produced at a unionized Bridgestone-Firestone plant is affected by a 3-year period of labor unrest; Bandiera et al. (2005) use personnel data from a fruit farm in the United Kingdom to analyze the impact of social preferences in the workplace among employees.

The evidence of the nonstandard behavior of employees—in the sense of the economic theory assumptions—raises the question of how rational agents should respond to the nonstandard behaviors of the others. DellaVigna (2009) discusses this question considering several fields: Industrial Organization, Labor Economics, Finance, Corporate Finance Political Economy, and Institutional Design.

Even before the recent interest in Behavioral Economics some authors were well aware of the noneconomic aspects to be taken into account when considering compensation and motivation. For instance, Baron and Kreps (1999) list and discuss the following aspects:

- Distributive and procedural justice: the perceived fairness of outcomes and of the processes by which the outcomes are determined influence the overall perception of what is fair in the workplace. According to Robbins and Judge (2009) distributive justice relates most strongly to the satisfaction of the outcomes and organizational commitment, whereas procedural justice is most strongly related to job satisfaction, employees trust, and citizen behavior.
- Social comparison: individuals' feelings about how fairly they are treated as compared to others. In particular Adams' (1965) equity theory is one of the most popular cognitive explanations of human behavior in work organizations.
- Social status: a socially defined position rank given to groups or group members by others; compensation should be consistent and even reinforce social status.
- Culture: the set of key values, assumptions, beliefs, understanding and norms that is shared by members of an organization (Daft and Noe 2001). Compensation policy should be consistent with the organization's culture, for example in an organization that promotes a familiar culture a strongly meritocratic incentive compensation system may be inappropriate.
- Intrinsic reward: refers to the satisfaction a person receives when performing a particular task or action or coming from the achievement of a goal. Sometimes extrinsic incentives can be counterproductive as in particular cases they may dull intrinsic motivation.

3 Incomplete Contracts

Although several contributions in contract theory assume that contracts are signed taking into account all variables that are or may become relevant—contract completeness—more recent developments analyze the effects of parties' inability to write contracts that take into account all possible contingencies.

Simon recognizes that contracts do not specify everything; in fact, he considers that "in an employment contract certain aspects of the worker's behavior are stipulated in the contract terms, certain other aspects are placed within the authority of the employer, and still other aspects are left to the worker's choice." (Simon 1951, p. 305).

This is well known in industrial/organizational psychology literature, Viteles (1932) argued that, although a company pays the same for labor, the outcome depends also on the employees working "with a will". Nevertheless, it took some time for this notion to be incorporated in Economics.

More recently, the important implications of incomplete contracting have been recognized both in terms of the efficiency of long-term economic relationships and as a possible explanation for the emergence of certain types of institutions such as the firm (see, e.g., Williamson 1985; and Klein et al. 1978).

In their 1988 seminal paper[2] Hart and Moore argue that the drafters of a contract face the difficult task of anticipating and dealing appropriately with the many contingencies which may arise during the course of their relationship. All things considered, as it may be prohibitively costly to specify the precise actions that each party should take in every conceivable eventuality, the contract which is written ends up being highly incomplete. As a consequence, Hart and Moore (1988) introduced the notion of contract incompleteness in order to take into account the impossibility for the parties to describe all of the states of the world in enough detail to later allow an outsider—for example a court—to verify which state had occurred. In other words, they acknowledge that it is impossible to sign a contract today that will be effective of all contingencies of a future date (Holmstrom and Tirole 1989).

Hart (1995, p. 23) argues that contracts may be incomplete as a consequence of three factors which are not considered in the standard principal-agent theory. First, given the complexity and unpredictability of the real word, it is difficult for individuals to predict all the contingencies which may arise in a future date. Second, even if it were possible to formulate individual plans it would be difficult for the contracting parties to find a common language to describe the states of the world and therefore to be able to negotiate these plans. Third, even if the parties could plan and negotiate the future contingencies, it would be difficult for them to write their plans in such a way that an outside authority can enforce them in the case of a dispute.

---

[2] Actually incomplete contracts are already considered in Grossman and Hart (1986).

According to Bolton and Dewatripont (2005), the introduction of incomplete contracts involves a substantive break. In fact, they argue, an incomplete-contracting perspective enables the focus on procedural and institutional-design issues instead of issues of compensation contingent on outcomes. Furthermore, another important issue raised by the incomplete-contract theory is the importance of the *ex ante* noncontractable actions. Maskin and Tirole (1999a, 1999b) have observed that, under certain conditions, *ex ante* non contractability of actions does not restrict implementability. Maskin and Tirole (1999a) argue that transaction costs—which usually lead to contractual incompleteness—need not to interfere with optimal contracting, provided that agents can probabilistically forecast at least their possible future payoffs. By contrast, Segal (1999) considers a situation in which even if contingencies cannot be described *ex ante*, parties cannot commit not to renegotiate, and only a finite number of actions can be described *ex post*, the first-best outcome cannot be achieved. Finally, Hart and Moore (1999), based on Segal's (1999) work, consider a hold-up problem and show that the first-best outcome may be unattainable even if states can be costlessly described *ex ante*.

The hold-up problem occurs in situations in which specific investments are observable by both parties but are nonverifiable and the cost of investment is born by the party who makes it. Although the investment will make both parties work more efficiently, they will refrain from doing so due to concerns that the investment may give the other party increased bargaining power, and thereby reduce their own profit. The hold-up problem is particular relevant to firms and often is exemplified by the 1920s relationship between Fisher bodies and General Motors (see, Klein et al. 1978). Nevertheless, it is also relevant to HRM as many investments—especially in human capital—are nonverifiable; therefore parties fearing to be expropriated of the surplus created by such specific investments tend to underinvest.

According to Gigerenzer and Todd (1999), in Simon's principle of bounded rationality there are two interlocking components: the structure of the environments in which the mind operates and the limitation of the human mind. Contract incompleteness takes these two aspects into account; yet even if bounded rationality in incomplete contracts has been discussed extensively (e.g., Hart 1990, and Tirole 2007, 2009), its role in modern organizational economics is still controversial; for a discussion see Foss (2001, 2003).

4 Contracts as Reference Points

Hart and Moore (2008) introduce an interesting behavioral hypothesis in contract theory. In their model they assume that an *ex ante* contract provides a reference point relative to which the parties evaluate *ex post* outcomes. In their model, a buyer and a seller meet on a competitive market before moving into a bilateral relationship. In the first stage they write an incomplete contract as there is uncertainty about the state of nature. In the second stage, the uncertainty is resolved and the parties observe the state but—in contrast to most of the existing literature—the trade does not become fully contractible *ex post*. To be more specific they assume that, although the broad outlines of *ex post* trade are contractible, the finer details are not. The change from the perfectly competitive market of the first stage to the bilateral monopoly is what Williamson (1985) terms a "fundamental transformation"; according to Hart and Moore (2008) potential candidates to explain the fundamental transformation are relationship specific investments or *ex post* search costs for alternative partners.

By assuming that *ex post* trade is only partially contractible Hart and Moore (2008) allow the seller to choose between two kinds of performance: perfunctory and consummate. The first one refers to performance within the letter of the contract; on the contrary consummate performance refers to performance within the spirit of the contract. In other words a gap remains between two levels of performance. Furthermore, they assume that whereas perfunctory performance can be judicially enforced consummate performance cannot. Finally, they suppose that a party provides consummate performance if he feels that he is getting what is he is entitled to, otherwise he will stint on the perfunctory performance. Hart and Moore (2008) use the term "shading" to indicate situations in which a party withholds some part of performance; they also assume that in terms of cost a party is completely indifferent between providing consummate and perfunctory performance.

In Fig. 1 we illustrate the three levels of complexity that make contracts incomplete. The larger triangle represents all the situations for which the parties may decide to try to write a contract. The left part of the triangle represents the situations for which it is impossible to predict all the contingencies which may arise in a future date. The right-hand triangle, by contrast, represents the situations for which the parties are able to predict all the contingencies. The right part of this triangle represents the situations in which the parties are able to predict the contingencies but are unable to negotiate them. The lower triangle represents the situations for which parties are able to predict all contingencies and to negotiate

Fig. 1 Levels of complexity for incomplete contracts

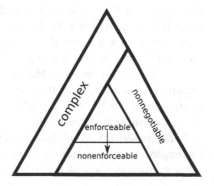

about. In the lower part of the triangle there are the situations for which the parties are able to predict the contingencies, are able to negotiate, but are unable to write their plans in such a way that an outside authority can enforce them in the case of a dispute. Finally, the central triangle represents the situations that are simple enough that the parties can write a complete contract. The small arrow represents the fact that the *ex ante* contract is just a reference point and the consummate performance cannot be judicially enforced.

It must be noted that both parties—the buyer and the seller—are allowed to shade. Although for the seller it seems more natural, being an example "working through the rules", also the buyer can shade, for example, stinting on what the seller needs to provide his performance.

Hart and Moore (2008) argue that the move from an *ex ante* competitive market to an *ex post* bilateral setting provides a rationale for the idea that contracts are reference points.

As the authors acknowledge there are several assumptions that need to be relaxed in this model, for example, the assumption that parties are interested only in the outcome and not in the process does not take into account the contributions of the procedural justice literature (Leventhal et al. 1980); this is an important point both in terms of retaliation on the workplace (Skarlicki and Folger 1997) and of organizational citizenship behavior (Moorman et al. 1998) which are important aspects related to counterproductive work behavior (Fox et al. 2001). Furthermore, even if the literature considers mainly relationships between two parties—the employee and the employer—justice has a huge impact on work group efficacy, see, e.g., Cropanzano and Schminke (2001). Since groups, rather than individuals, are the fundamental unit of work in modern organizations (Finholt and Sproull 1990) the importance of this topic cannot be overlooked.

Some experimental evidence about Hart and Moore's (2008) model is provided in (Fehr et al. 2011). They tested the main features of the model in a laboratory experiment and are able to confirm the empirical relevance of the behavioral forces described in Hart and Moore (2008) when a fundamental transformation takes place.

Furthermore, in Fehr et al. (2009), experimental analysis is extended to understand the role of fundamental transformation. Their laboratory evidence shows that, in the absence of a fundamental transformation, contracts no longer provide reference points as the sellers no longer perceive the contracts in these terms. In this case, since the contract loses this role other variables need to be considered in order to fill this gap.

Nevertheless, Hart and Moore (2008) bring a totally new perspective in incomplete contracts literature; in fact, in their view, a contract provides a reference point for the trading relationship and what the parties feel entitled to. Also, the entitlement ensued by the contract limits disagreement, aggrievement, and the consequences of shading. Their model sheds lights on how the gaps of incomplete contracts may be filled either by courts or—as we will suggest—by other constructs such as the psychological contract.

5 Psychological Contract

In the Industrial/Organizational literature, it is well known that successful firms depend on workers who voluntarily cooperate and that formal control systems are not able to compel critical workers contributions (Rousseau 2010). Evidence of this problem is well known, for example Smith (1977) provides empirical evidence showing how employees who are more satisfied with their job are more likely to exert a discretionary effort, or—using Hart and Moore terminology—to provide a consummate performance instead of a perfunctory one.

The interest in the construct of the psychological contract by I/O psychologists arises from the necessity to face this kind of dilemma central to the individual-organization exchange relationship.

The term "psychological contract", was first introduced by Argyris (1960) to describe the relationship between factory line employees and their foremen; then Levinson (1962) expanded its definition and Schein (1965) provided a different perspective.

Later, Kotter (1973) defined it as "an implicit contract between an individual and his organization which specifies what each expect to give and receive from each other in their relationship".

Rousseau's (1989) contribution was a transition point; in her paper, the author defined a psychological contract as "an individual's belief regarding the terms and conditions of a reciprocal exchange agreement between that focal person and another party" (p. 123). Furthermore—she continues—a psychological contract emerges when an individual perceives that contributions he or she makes obligate the organization to reciprocity or vice versa. Therefore—by definition—a psychological contract is the perception of an exchange between oneself and another party (Rousseau 1998).[3]

Some critical aspects of the psychological contract have been pointed out especially when research departed from the collective conceptualization of a joint "psychological contract" between the two parties (see for example Guest 1998). These aspects have been discussed in Rousseau (1998). Furthermore, integrating the individual level perspective—in terms of the individual's system of beliefs—has provided a better understanding of the separate and joint effects of psychological contracts (Rousseau 2010).

In terms of incompleteness, there is an interesting symmetry between the psychological contract and the recent developments of contract theory we discussed in the previous section. In fact, as Rousseau (2010) suggests, at hire also psychological contracts suffer from the incapacity of the parties to spell out all of the details of the relationship. Although the reasons for incompleteness of the psychological contract are similar to those considered in Hart (1995) and discussed

---

[3] The reader interested in a historical analysis of the psychological contract may refer to Roehling (1997); Rousseau (2010) provides also an analysis of the cognitive, emotional, and behavioral processes underlying it.

in the previous sections, psychological contracts evolve over time as new demands modify the relationship. Another important point is that psychological contracts vary across firms and workers. According to Rousseau (2010) they can vary from considering only economic terms—and in this case are quite similar to contracts considered in the economic theory—to being extremely complex, as in the case of high-involvement work.

Two aspects of the psychological contract are important especially when compared to the assumptions of economic theory. First, according to Rousseau (2010), a stable contract promotes goal-oriented behaviors and does not require heavy control and monitoring from the employee. Second, the process by which a psychological contract is transformed must be carefully monitored to avoid the risk of contract breach and violation.

6 Putting the Pieces Together

As we have seen, both Economics and Psychology agree—even if from slightly different perspectives—that contracts are not complete. In fact, whereas Economics acknowledges incompleteness as the result of the complexity of the world, and analyzes the consequence of such an incompleteness from the point of view of renegotiation costs (Hart 1995), in Psychology the psychological contract is what provides the fleshing out of the otherwise incomplete arrangements between employees and firm (Rousseau 2010). Both disciplines agree on the impossibility of spelling out the details of an employment relationship: "changing circumstances mean that not all contingencies can be foreseen" (Rousseau 2010, p. 198) and "it is hard for the contracting parties to *negotiate* these plans, not least because they have to find a common language to describe states of the world and actions with respect to which prior experience may not provide much of a guide. [...] even if the parties can plan and negotiate the future, it may be very difficult for them to write the plans down in such a way that in the event of a dispute, an outside authority -a court, say, can figure out what these plans mean and enforce them" (Hart 1995, p. 23).

Furthermore, even if psychological contracts tend to be incomplete at the beginning of the outset and tend to evolve over time as the relationship develops (Rousseau 2010) their evolution may allow the parties to adapt to the new contingencies. By adding the perspective of the psychological contract to economic analysis of contracts it is possible to obtain a new dimension to HRM. Indeed we can think that when the psychological contract considers only the economic aspects of the relationship it may be considered as equivalent to an economic contract as illustrated in Fig. 2a). In this case it covers more or less the simpler situations which can be formalized by complete contracts. On the contrary, when the relationship evolves and the psychological contract expands as the result of the new contingencies, it may become a tool to cover also situations which are either nonenforceable or non negotiable or even more complex ones. This aspect is

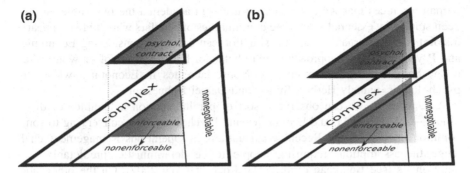

Fig. 2 **a** The psychological contract at the beginning, **b** as it expands over time

related to time and dynamics processes; it may be also referred to what Binmore (1992) calls adjusting to circumstances.[4] In fact two aspects that are central to our analysis are bounded rationality and dynamical processes. The first aspect clearly refers to the incapacity of parties to write a complete contract and the other to evolution of psychological contracts (Rousseau 2010).

As we have seen, Hart and Moore (2008) introduce the assumption that *ex ante* contracts provide reference points for entitlements in *ex post* trade. This approach is not only interesting per-se but especially when contrasted with the literature on psychological contract. In fact, at the outset of the employment both the psychological and formal contracts become the core and the reference point around which the working relationship is built.

According to Loch and Wu (2007) "it is evident that the normative models based on hyper-rationality assumptions, which are popular in OM research, may not help much in the complex reality". The same authors also observe how "Camerer's (1999) diagnosis of incompleteness of economics and both its complementarity with and separatedness from psychology closely parallels the history of Operations Management and Organizational Behavior". Considering the fundamental cognitive processes operating on the thinking and behavior of the parties may provide a partial solution to the issues arising from the complexity of the world. In this sense, the psychological contract complements the approach used in Economics as it considers the mental operations involving emotional and non-emotional processes, which are otherwise neglected.

7 Conclusion

Disciplines with a long history of separateness are studying the same problems from different perspectives. Examples are OM and HRM. As Boudreau et al. (2003) observe, in industry it has been rare for an operations manager to become a

---

[4] I am grateful to Gian-Italo Bischi for providing this insight.

human resources manager, or vice versa, and in academia the two subjects have been studied by essentially separate communities of scholars who publish in nearly disjoint sets of journals. A similar situation can be found considering Economics and Psychology. Both approach human behavior but the points of view are often polar. Some notable exceptions are Nobel laureates in Economics who were psychologists, namely Herbert Simon and Daniel Kahneman.

Yet, recently, we can observe a sort of parallel motion as couples of disciplines—OM and HRM, and Economics and Psychology—are converging to some common interest topics. Recent contributions in Operations Management challenge the assumptions requiring people to be deterministic, predictable, and emotionless (see Boudreau et al. 2003; Loch and Wu 2007). On the other side, Behavioral Economics challenges and relaxes some of the neoclassical assumptions on individuals, mainly the one according to which people are self-interested rational agents with stable preferences.

In my opinion, a promising area for a multiple perspective approach is work relationships. Simon's (1951) analysis outlined "the assumptions of rational utility-maximizing behavior incorporated in it" as one of the main limitations of the current view of labor management. In the following years several authors provided interesting contributions on these topics; nevertheless the hyper-rationality assumptions are still popular both in OM (Loch and Wu 2007) and in Economics (Bolton and Dewatripont 2005).

Both formal contracts—in the sense of Economics—and psychological contracts are instruments useful to obtain the critical commitment which may make the difference between the success or failure of a firm. Contract incompleteness has been recognized in Economics; this contribution shows how psychological contracts may be interpreted as a way to fill the gaps of contracts. In particular, psychological contracts fill these gaps considering human behavior in organizations by building upon fundamental processes that are not usually considered in Economics. Given the recent interest Economics has in considering behaviors which go beyond the *Homo Oeconomicus* paradigm and recent contributions of Behavioral Operation Management these aspects, seem to be relevant to approach these important issues.

The road is long and a first suggestive hypothesis is to see psychological contract and contract theory as a way to gain employees' commitment. Although these two approaches are for the most self-centered on the disciplines they belong to a wider scope approach which is in order. In this way we can avoid being trapped in the assumptions that characterize the different disciplines.

Finally, it must be mentioned that in this analysis we did not consider the useful contributions made by scholars of other disciplines. For example, contract incompleteness has been considered in the Industrial Relations literature by other authors such as Marsden (2004) and also by Collins (2001, p. 23) who argued that "the model of the employment contract that fits into this traditional scheme is the contract that is incomplete by design".

Adding the perspective coming from other fields that have extensively been working on contracts may add new insight. In particular, integrating and articulating psychological, economic/incentive, and legal aspects of contracts in terms of contract breach and renegotiation can suggest how to approach the complexity of working relationships.

References

R.L. Ackoff, Science in the systems age: beyond IE, OR, and MS. Oper. Res. **21**(3), 661–671 (1973)

J.S. Adams, Inequity in social exchange, in *Advances in Experimental Social Psychology*, ed. by L. Berkowitz (Academic Press, New York, 1965), pp. 276–299

C. Argyris, *Understanding Organizational Behavior* (Dorsey Press, Homewood, 1960)

O. Bandiera, I. Barankay, I. Rasul, Social preferences and the response to incentives: evidence from personnel data. Quart. J. Econ. **120**(3), 917–962 (2005)

J.N. Baron, D.M. Kreps, *Strategic Human Resources* (Wiley and Sons, New York, 1999)

K. Binmore, *Fun and Games. A Text on Game Theory* (Heath and Co, Lexington, 1992)

R. Birnbaum, The life cycle of academic management fads. J. High. Educ. **71**(1), 1–16 (2000)

P. Bolton, M. Dewatripont, *Contract Theory* (The MIT Press, Cambridge, 2005)

J. Boudreau, W. Hopp, J.O. McClain, L.J. Thomas, On the interface between operations and human resources management. Manuf. Serv. Oper. Manag. **5**(3), 179–202 (2003)

C. Camerer, Behavioral economics: reunifying psychology and economics, in *Proceedings of the National Academy of Science USA*, vol. 96 (USA, 1999), pp. 10575–10577

P.P. Carson, P.A. Lanier, K.D. Carson, B.N. Guidry, Clearing a path through the management fashion jungle: some preliminary trailblazing. Acad. Manag. J. **43**(6), 1143–1158 (2000)

G. Charness, M. Rabin, Understanding social preferences with simple tests. Quart. J. Econ. **117**(3), 817–869 (2002)

R.H. Coase, The nature of the firm. Economica **4**(16), 386–405 (1937)

H. Collins, Regulating the employment relationship for competitiveness. Indus. Law J. **30**(1), 17–47 (2001)

R. Cropanzano, M. Schminke, Using social justice to build effective work groups, in *Groups at Work: Advances in Theory and Research*, ed. by M. Turner (Erlbaum, Hillsdale, 2001), pp. 143–171

R.L. Daft, R.A. Noe, *Organizational Behavior* (South-Western Publishing, Mason, 2001)

G.J. DeAngelo, S. Beckman, L. Chen, J.W. Smith, X. Zhang, *Microeconomics and psychology*. SSRN, http://ssrn.com/abstract=1893788. Accessed 23 July 2011

S. DellaVigna, Psychology and economics: evidence from the field. J. Econ. Lit. **47**(2), 315–372 (2009)

G. Divotti, U. Merlone, in *Envy and Money Burning: Some Contradictory Evidence*. Paper presented at the envy at work international symposium (Torino, Italy, 2011)

E. Fehr, S. Gächter, Fairness and retaliation: the economics of reciprocity. J. Econ. Perspect. **14**(3), 159–181 (2000)

E. Fehr, O. Hart, C. Zehnder, Contracts, reference points, and competition-behavioral effects of the fundamental transformation. J. Eur. Econ. Assoc. **7**(2–3), 561–572 (2009)

E. Fehr, O. Hart, C. Zehnder, Contracts as reference points—experimental evidence. Am. Econ. Rev. **101**, 493–525 (2011)

T. Finholt, L.S. Sproull, Electronic groups at work. Organ. Sci. **1**(1), 41–64 (1990)

N. Foss, Bounded rationality in the economics of organization: present use and (some) future possibilities. J. Manage. Governance **5**(3–4), 401–425 (2001)

N. Foss, Bounded rationality in the economics of organization: much cited and little used. J. Econ. Psychol. **24**(2), 245–264 (2003)

S. Fox, P.E. Spector, D. Miles, Counterproductive work behavior (CWB) in response to job stressors and organizational justice: some mediator and moderator tests for autonomy and emotions. J. Vocat. Behav. **59**(3), 291–309 (2001)

X. Gabaix, D. Laibson, G. Moloche, S. Weinberg, Costly information acquisition: experimental analysis of a boundedly rational model. Am. Econ. Rev. **96**(4), 1043–1068 (2006)

J. Geanokoplos (2004), The Arrow-Debreau model of general equilibrium. (Cowles Foundation Paper 1090). Yale University, New-Haven, CT, http://cowles.econ.yale.edu/P/cp/p10b/p1090.pdf. Accessed 14 Feb 2012

G. Gigerenzer, P.M. Todd, Fast and frugal heuristics, in *Simple Heuristics that make us Smart*, ed. by G. Gigerenzer, P.M. Todd, ABC Research Group (Oxford University Press, Oxford, 1999), pp. 3–34)

S.J. Grossman, O.D. Hart, The costs and benefits of ownership: a theory of vertical and lateral integration. J. Polit. Econ. **94**(4), 691–719 (1986)

D.E. Guest, Human resources management and industrial relations. J. Manage. Stud. **24**(5), 503–521 (1987)

D.E. Guest, Is the psychological contract worth taking seriously? J. Organ. Behav. **19**, 649–664 (1998)

O.D. Hart, Is "bounded rationality" an important element of a theory of institutions? J. Inst. Theor. Econ. **146**(4), 696–702 (1990)

O.D. Hart, *Firms contracts and financial structure* (Oxford University Press, Oxford, 1995)

O.D. Hart, J. Moore, Incomplete contracts and renegotiation. Econometrica **56**(4), 755–785 (1988)

O.D. Hart, J. Moore, Foundations of incomplete contracts. Rev. Econ. Stud. **66**(1), 115–138 (1999)

O. D. Hart, J. Moore, Contracts as reference points. *Q. J. Econ.* **CXXIII**(1), 1–48 (2008)

B. Holmström, J. Tirole, The theory of the firm, in *Handbook of Industrial Organization*, ed. by R. Schmalensee, R.D. Willig, vol. I (North-Holland, Amsterdam, 1989), pp. 61–133

B. Klein, R. Crawford, A. Alchian, Vertical integration appropriable rents, and the competitive contracting process. J. Law Econ. **21**, 297–326 (1978)

J.P. Kotter, The psychological contract. Calif. Manage. Rev. **15**, 91–99 (1973)

A.B. Krueger, A. Mas, Strikes, scabs, and tread separations: labor strife and the production of defective bridgestone/firestone tires. J. Polit. Econ. **112**(2), 253–289 (2004)

J.J. Laffont, D. Martimort, *The Theory of Incentives* (Princeton University Press, Princeton, 2002)

K. Legge, *Human Resource Management. Rhetorics and Realities*, Anniversary edn. (Palgrave MacMillan, Basingstioke, 2005)

G.S. Leventhal, J. Karuza Jr., W.R. Fry, Beyond fairness: a theory of allocation preferences, in *Justice and Social Interaction*, ed. by G. Mikula. Experimentation and Theoretical Research from Psychological Research (Plenum, New York, 1980), pp. 167–218

H. Levinson, *Organizational Diagnosis* (Harvard Univ. Press, Cambridge, 1962)

C.H. Loch, Y. Wu, Behavioral operations management. *Found. Trends Technol. Inf. Oper. Manag.* **1**(3), 121–232 (2007). http://dx.doi.org/10.1561/0200000009

G. Loewenstein, J.S Lerner, *The role of affect in decision making*. ed. by R. Davidson, H. Goldsmith, K. Scherer, Handbook of Affective Science. (Oxford: Oxford University Press, 2003), pp. 619–642

M.S. Logan, Using agency theory to design successful outsourcing relationships. Int. J. Logist. Manag. **11**(2), 21–32 (2000)

I. Macho-Stadler, D. Pérez-Castrillo, *An Introduction to the Economics of Information* (Oxford University Press, New York, 1997)

D. Marsden, The network economy and models of the employment contract. Br. J. Indus. Relat. **4**(4), 659–684 (2004)

A. Mas, Pay, reference points, and police performance. Quart. J. Econ. **121**(3), 783–821 (2006)

E. Maskin, J. Tirole, Unforeseen contingencies and incomplete contracts. Rev. Econ. Stud. **66**, 83–114 (1999a)

E. Maskin, J. Tirole, Two remarks on the property-right literature. Rev. Econ. Stud. **66**, 139–149 (1999b)

P.R. Milgrom, J. Roberts, *Economics Organization and Management* (Prentice Hall, Englewood Cliffs, 1992)

R.H. Moorman, G.L. Blakely, B.P. Niehoff, Does perceived organizational support mediate the relationship between procedural justice and organizational citizenship behavior? Acad. Manag. J. **41**(3), 351–357 (1998)

S.P. Robbins, T.A. Judge, *Organizational Behavior*, 13th edn. (Pearson Education, London, 2009)

M.V. Roehling, The origins and early development of the psychological contract construct. J. Manag. Hist Arch. **3**(2), 204–217 (1997)

D.M. Rousseau, Psychological and implied contracts in organizations. Empl. Responsib. Rights J. **2**(2), 121–139 (1989)

D.M. Rousseau, The problem of psychological contract considered. J. Organ. Behav. **19**, 665–671 (1998)

D.M. Rousseau, The individual-organization relationship: the psychological contract, in *APA Handbook of Industrial and Organizational Psychology*, vol. 3, Maintaining, Expanding, and Contracting the Organization, ed. by S. Zedeck (American Psychological Association, Washington, DC, 2010), pp. 191–210

B. Salanie, *The Economics of contracts. A Primer* (The MIT Press, Cambridge, 1997)

E.H. Schein, *Organizational Psychology* (Prentice Hall, Englewood Cliffs, 1965)

R.G. Schroeder, *Operations Management: Decision Making in the Operations Function*, 4th edn. (McGraw-Hill, Singapore, 1993)

I. Segal, Complexity and renegotiation: a foundation for incomplete contracts. Rev. Econ. Stud. **66**(1), 57–82 (1999)

H.A. Simon, A formal theory of the employment relationship. Econometrica **19**(3), 293–305 (1951)

H.A. Simon, *Models of Man* (John Wiley & Sons, New York, 1957)

H.A. Simon, *Administrative Behavior*, 4th edn. (Free Press, New York, 1997)

F.J. Smith, Work attitudes as predictors of attendance on a specific day. J. Appl. Psychol. **62**, 16–19 (1977)

V.L. Smith, Review: rational choice: the contrast between economics and psychology. J. Polit. Econ. **99**(4), 877–897 (1991)

D.P. Skarlicki, R. Folger, Retaliation in the workplace: the roles of distributive, procedural, and interactional justice. J. Appl. Psychol. **82**(3), 434–443 (1997)

J. Tirole, Bounded rationality and incomplete contracts (2007). University of Toulouse,. http://citeseerx.ist.psu.edu/viewdoc/download?doi=10.1.1.123.5630&rep=rep1&type=pdf

J. Tirole, Cognition and incomplete contracts. Am. Econ. Rev. **99**(1), 265–294 (2009)

M.S. Viteles, *Industrial Psychology* (Norton, New York, 1932)

O. Williamson, *The Economic Institutions of Capitalism* (Free Press, New York, 1985)

J. Zizzo, A.J. Oswald, Are people willing to pay to reduce others' incomes? Annales d'Économie et de Statistique **63**(64), 39–65 (2001)

Chapter 8
Incentives for Cost Transparency Implementation: A Framework from an Action Research

Pietro Romano and Marco Formentini

Abstract The aim of this research is to develop a framework to support cost transparency implementation. Though cost transparency is a well-known practice in supply chain management literature, there is a lack of guidelines supporting managers to effectively implement it. The study has been developed using empirical findings from an action research, the authors conducted in close collaboration with an Italian manufacturer of modular kitchens and 19 suppliers. This research discusses a methodology to support (1) the selection of those customer–supplier relationships worth being developed into cost transparency, and (2) the identification, for each of these relationships, of appropriate forms of actions/ incentives to stimulate suppliers to share cost information. The key outcome lies in the theoretical framework emerging from on-field empirical evidences supporting managers in cost transparency implementation. The proposed model fills some gaps found in supply chain literature and explicitly addresses an unusual variable to be considered in the suppliers' selection phase: the buyer's interest. From a managerial point of view our model provides a framework buying companies use to select and classify the subset of suppliers willing and worth being engaged in cost transparency initiatives, and to identify appropriate actions to implement cost transparency. The findings of a single action research cannot be generalized to the overall population of manufacturing entities. Large sample data collection efforts

P. Romano (✉)
Department of Electrical Managerial and Mechanical Engineering,
University of Udine, Via Delle Scienze 208 33100 Udine, Italy
e-mail: pietro.romano@uniud.it

M. Formentini
Faculty of Management—Cass Business School, City University London,
106 Bunhill Row, London EC1Y 8TZ, UK
e-mail: Marco.Formentini.1@city.ac.uk

I. Giannoccaro (ed.), *Behavioral Issues in Operations Management*,
DOI: 10.1007/978-1-4471-4878-4_8, © Springer-Verlag London 2013

will be needed to test the proposed findings and make our prescriptions more robust. Additionally, the proposed model could be tested in other industries outside the modular kitchen industry.

1 Introduction

The concept of cost transparency (CT) concerns the flow of cost information between companies within the supply chain. The aim is to develop mutual commitment in customer–suppliers dyads, to acquire knowledge about upstream or downstream processes and to conduct joint activities with supply chain partners to reduce costs. Although there is a growing interest in the cost information sharing field (Kajüter and Kulmala 2005; Agndal and Nilsson 2008), recent studies lament the lack of empirical evidence regarding how to implement CT practices and how to avoid pitfalls that could vanish managerial efforts (Chin-Chun et al. 2008; Piontkowski and Hoffjan 2008).

To fill this gap in literature we conducted an action research analyzing the relationships between an Italian manufacturer of modular kitchens and fifteen of its suppliers. This study develops a model to select within the buying company supplier portfolio those relationships worth being developed into CT and to identify what actions/incentives toward cost information sharing need to be activated according to the characteristics of customer–supplier relationships.

The paper is organized as follows: in the next section we provide the literature review, from which emerges the lack of guidelines to implement CT. The third section describes the research design and the trial and error process that allowed the development of the proposed CT implementation model. The following sections describe the proposed model and discuss the results. In the conclusions, we present the academic and managerial implications together with some hints for future research.

2 Theoretical Background

2.1 Cost Transparency

One of the key issues in supply chain management literature is the need for managers in an organization to know more about what takes place in other firms—especially their counterparts in the upstream network. Pressure on prices and costs from fierce international competition, the trend toward outsourcing and resource scarcity are forcing managers to explore new ways to obtain sensitive—or even secret—information and knowledge from their suppliers, without having to resort to acquisition. This fact implies the need for many companies, along with an ever more careful analysis of their internal costs, to monitor the costs of their suppliers and possibly the entire supply chain. The exchange of cost information is a strong feature of the lean supply model, where CT is defined as "the sharing of costing information between customer and supplier including data which would traditionally be kept secret by each party, for use in negotiations. The purpose of this is to make it possible for customer and supplier to work together to reduce costs" (Lamming 1993).

Several authors dealt with the exchange of cost information between manufacturers and assemblers and introduced the terms of *open book accounting* (Mouritsen et al. 2001; Kulmala et al. 2002), *open book negotiation* (Lamming et al. 2005), *open book costing* (McIvor 2001), *open books policy* or more generally *open books* (Agndal and Nilsson 2008). In particular, CT plays a key role in inter-organizational cost management literature (Cooper and Slagmulder 1999; Mouritsen et al. 2001; Kajüter and Kulmala 2005). In this setting, cost management identifies methods and techniques to monitor and manage costs date within companies. The goal of inter-organizational cost management is to create synergies among such methods and techniques to reduce costs through the entire supply chain. A high level of coordination between all the companies within the supply chain is required to achieve these synergies. Interestingly enough, Seal et al. (1999) maintains that inter-organizational cost control becomes even more important when the duration of the customer–supplier relationship increases.

Inter-organizational cost management literature recognizes that the main purpose of CT is to stimulate customer and supplier to work together to eliminate wastes at their interface and to capture value for both participants (Agndal and Nilsson 2008). Therefore, we can argue that CT should not be seen as a cost reduction tool only, as it can be effectively used to improve relations among counterparts in supply networks. In other words, sharing confidential cost information may lead to an increase in the level of trust, cooperation, and commitment between the parties (Kulmala 2004; Agndal and Nilsson 2008). Programs that improve mutual trust may also bring significant benefits to the supplier in terms of increased sales volumes and margins, allowing it to become a preferred supplier. Kulmala (2004) maintains that the level of trust between companies, besides being a requirement for implementing open book accounting is a direct consequence of the reduction information asymmetry between the parties. The author underlines that cost information exchange is essential in order to develop win–win relationships that involve, for example, the sharing of profits. Opening the books can also be a way to ease the tension occasionally occurring in negotiations particularly in regard to pricing (Agndal and Nilsson 2008).

2.2 Main Problems in CT Implementation

The main concern about CT is tied to the opportunistic use of cost information by the buying company. In this case the supplier notices that the nature of the open books is an attempt of the customer to lower prices by exercising a form of contractual power.

To make managers aware of this problem, Lamming et al. (2005) introduce the
concept of "value transparency", in contrast to what they define "one way open
book negotiation", which is limited to the unidirectional information exchange from
suppliers to customers, without sharing benefits with suppliers. Value transparency
implies the mutual engagement pursuing the aim to reduce costs and all the causes of
value wastes in both customer and supplier activities. In order to reduce the possibility of opportunistic behaviors, the information granted to the customer would
have to only regard the directly interested areas and aspects of the improvement
plans.

Kajüter and Kulmala (2005) identify a list of reasons of failure in open book
accounting implementation:

- Suppliers experience no extra-benefit from openness and main contractors do
  not offer win–win solutions.
- Suppliers think that accounting information should be kept in-house.
- Network member cannot produce accurate cost information and see no sense in
  sharing poor cost data.
- Suppliers are afraid of being exploited if they reveal their cost structure.
- Suppliers do not have capable resources or resource support from main contractors
  for the development of accounting systems.
- Network members cannot agree on how open-book practice should be implemented.

In addition, McIvor (2001), in his analysis of failures in cost reduction programs
based on information exchanges, points out that the culture of "people" at both the
customer and supplier interface is a considerable barrier to CT implementation.

Cost information sharing seems to be accepted by suppliers only if it is perceived as an attempt by the customer to improve the upstream processes
(Lamming 1996). In order to initiate a fair exchange of information on costs it is
necessary to plan joint collaborations between buying companies and suppliers,
based on shared risks and benefits. In this sense, a pivotal aspect addressed by all
the studies on CT we analyzed is the need for a "two-way" flow of information.
When the buyer requests the supplier to share information on its products or
processes, also information on the buyer's products or processes should be shared
to commit the supplier toward cooperation and, therefore, to implement CT.

2.3 Frameworks for CT Implementation

Literature has well described on the concept of CT, its fundamental difference
from established practices such as one-way open book negotiation, its beneficial
effects for both customers and suppliers, and its fit with some other management
techniques such as target costing, ABC costing, and kaizen improvement
(Lamming et al. 2005; Agndal and Nilsson 2008). The main under investigated
issue we have identified in literature concerns how to implement CT. This is a
practical matter that has a non-negligible theoretical implication.

In literature we have found two frameworks, developed by Kulmala (2004) and Kajüter and Kulmala (2005), investigating how some factors impacts the need and the willingness of companies to launch open book accounting programs with a CT perspective. However, these frameworks do not lead to models or guidelines that can support buying companies in the implementation of regular exchange of cost information with suppliers.

In fact, the first contribution (Kulmala 2004) suggests a three-dimensional (3D) framework that considers the balance of power within the relationship (split in customer dominant or supplier dominant), trust (split in adequate or not adequate) and the level of mutual business (split in high or low), and aims to provide to the assembler a first help to decide whether the possible implementation of an open book accounting program can succeed or not. Interestingly enough, this framework does not explicitly consider the convenience for the buying company to invest on a CT program with certain suppliers.

The second study (Kajüter and Kulmala 2005) develops a contingent model which identifies the variables impacting on CT implementation. There are three types of context factors that induce the disclosure of cost data in networks: exogenous environmental factors (degree of competition, economic trends), network specific factors (type of network, type of product, infrastructure, social nature of network relationships), and endogenous firm-specific factors (firm size, cost accounting systems, competitive policy, commitment). However, from the buyer's point of view, it is not clear how managers can select suppliers with whom to start collaboration for knowledge sharing on costs. Moreover, once identified such suppliers, there is not enough theoretical support to help managers stimulate counterparts to share cost information.

2.4 Research Questions

We posit this chapter in the stream of literature on supply chain management implementation, as it aims to fill the gaps discussed above by providing a model to address the following research questions.

RQ1. From the literature emerges a key factor for the success of CT implementation lies in the identification of those suppliers to involve in cost transparency programs and in the elimination of all those counterparts which would not guarantee the achievement of the desired improvements. Most companies use not to share cost information simply because they fear to lose their know-how and/or to see reduced their bargaining power. Therefore, it is necessary to understand what are the key variables that preclude suppliers' participation in any collaborative form based on CT. The first research question can be specified as: *how can buying companies identify those customer–supplier relationships worth being developed into CT?*

RQ2. As proposed by IMP group, firms interact in a relationship "atmosphere" (Håkansson 1982), that is described in terms of the degree of mutual expectations held by firms in a business relationship and is characterized by the states of closeness/

distance, conflict/cooperation, and power/dependence. The buying company should be able to identify what actions/incentives for CT implementation to activate depending on the specific "atmosphere" characterizing each relationship with suppliers. Therefore, the second research question is: *how can buying companies identify what forms of actions/incentives toward CT to activate according to the specific "atmosphere" characterizing relationships with suppliers?*

3 The Action Research

This study is an action research which entailed the active participation of the researchers in a firm's customer–supplier relationship management process. According to Yin (2003), qualitative research methods are suited to answer "how"', and "why" open questions like our research questions. Furthermore, action research helps to address research objectives within a natural setting, which would otherwise be expensive, difficult, or impossible to replicate and also difficult to be expressed by statistical analyses. Näslund (2002) argues that action research is particularly appropriate to developing research within an applied field such as supply chain management. In addition, action research is a suited approach when studying the implementation process as in this study, because it allows researchers to spend time directly in organizations, to gather first-hand information, and to develop concrete knowledge through continuous interactions with the "real world" (Coughlan and Coghlan 2002).

As usually happens in action research, the model we developed emerged from iterative cycles of data gathering, feedback, analysis, action planning, implementation, and evaluation.

3.1 Research Context

The objects of analysis are the customer–supplier relationships between an Italian manufacturer of modular kitchens and fifteen of its suppliers. The buyer company (268 million € turnover; among the five largest kitchen manufacturers in Europe; 1,650 employees in eight production plants located in Italy, France, and Germany) purchases the major part of the components from external suppliers, and successively assembles the final product. Therefore, the buyer company needs to continuously control the price level of the purchased products and the impact of variations of costs of raw materials and other inputs. For this reason the firm has developed a cost estimation software, aimed to support cost analysis and benchmarking in the purchasing phase and also in the new product development process.

At the beginning of this study, we were asked to identify mechanisms facilitating cost information exchanges between the buying company and its suppliers, and to support the assembler in the selection of actions/incentives to stimulate suppliers to feed with cost information in their possession the cost estimation software.

Due to these features, we found the action research methodology suited to our investigation. In fact, our research goals fit the major characteristics of action research as defined by Westbrook (1995): to solve a firm problem (i.e., creating a structured and formalized cost information-sharing flow with suppliers) and to make a contribution to scientific knowledge and theory, addressing to gaps found in literature.

3.2 Research Process

The results of this research are the outcomes of an iterative action research process, which has been developed in continuous interaction with the buying company and its suppliers. The research project lasted 1 year, from February 2008 to 2009. We worked actively on the project in order to prior understand the context of our study, then gather and analyze data (which came primarily from four sources: semi-structured interviews, internal documents, software applications, and direct observations), and finally plan, implement, and evaluate necessary actions. These activities were cycled to warrant continuous adjustment to new information and events, while a monitoring step took place over time in the form of scheduled team meetings or informal interviews or observations.

Several meeting have been organized with purchasing managers and operations managers of the buying company and its suppliers with the aim to maintain a tight relationship between the members of the research team. Moreover, all the members have always been informed about the evolution of the study thanks to the draft of biweekly reports.

As exhibited in Fig. 1, in conducting the action research we followed the iterative cyclical process model proposed by McKay and Marshall (2001). This process enabled us to collect very detailed data and to gain in-depth insights by closely interacting with the modular kitchen company and the selected suppliers. At the same time it allowed the researchers to alternate periods of data collection and periods of reflection away from the company. Each research loop was constantly supported by literature review, data analyses, and meetings.

To present the action research results, we first describe the trial and error process that allowed us to develop our CT implementation model, then we describe and discuss the framework.

3.3 First Action Research Loop

Action research involves the continuous redefinition of the research objectives. Initially, the research process aimed to diagnose problems related to strengthening

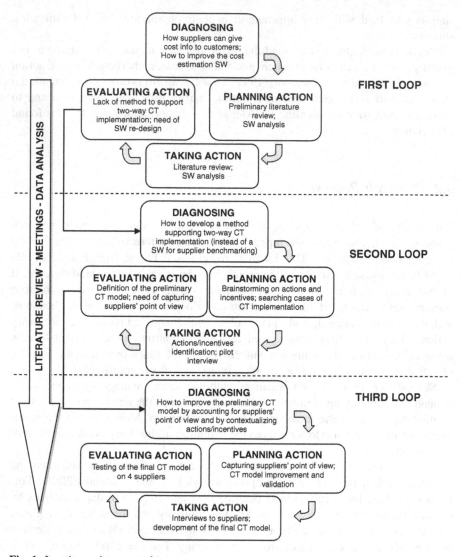

Fig. 1 Iterative action research process

cost information exchanges with suppliers and to improve the cost estimation and benchmarking software already used by the company, if necessary also exploring the opportunity to design a new tool. To achieve such objectives, researchers planned and performed a careful review of organizational literature, focusing on the search of a suitable methodology to involve suppliers in the exchange of cost information with customers. In particular, the research team found interesting hints in the concept of CT developed in the lean supply model by Lamming (1993). However, this preliminary literature review discovered also the lack of a consolidated methodology to effectively implement CT. In the literature, we found only scarce materials describing how to successfully involve suppliers in CT programs.

CT literature underlines the need to use information gained from suppliers in cost reduction programs and to fairly share with the suppliers the benefits. Alternatively, the only way to "convince" suppliers to share information is to lever on a higher contractual power, thus realizing the one-way open book negotiation.

During the first loop, researchers also planned to carefully analyze the existing cost estimation software. It soon emerged the need to interview suppliers, in order to understand (1) their attitude toward feeding the software with cost information in their possession and (2) the main limits of the software in dealing with different categories of purchased components.

At the same time, another critical issue emerged. While literature alerts from several pitfalls thwarting CT programs, it does not provide any help to avoid the most common error in its implementation, namely the use of cost information to achieve supplier benchmarking only. In fact, the buying company purchasing manager himself confessed that it is unfair claiming to involve a supplier in cost data sharing in order to successively use the gained information against him in the negotiation phase. Hence, both suppliers and the buying company must perceive the cost estimation software not as a benchmarking tool, but as a device to highlight areas where it is possible to jointly plan cost reduction activities. Therefore, the initial research objective has been redefined with a wider scope: to develop a model aimed not only at supporting negotiation from the buying company's point of view, but also at achieving the continuous improvement of the relations with suppliers.

3.4 Second Action Research Loop

After the first research loop the study aim has been reshaped to address the search and development of guidelines to effectively involve suppliers in "two-way" information to exchange programs promoted by the buying company and finalized to joint cost reduction, respecting suppliers' know-how, and sharing the benefits adopting a "win–win" perspective. This change in the research aim determined a change in the role of the cost estimation software, as it becomes a fundamental tool to identify what suppliers are better disposed to collaborate and what purchased components are worth to be addressed in CT programs. Moreover, at this stage, the buying company realized that it was strategic to better identify the kind of information on costs to be exchanged and the forms of action/incentives toward information exchanges. The goal for the action research team was to develop a preliminary methodology to effectively implement two-way CT.

Researchers came back to literature and generated several ideas to stimulate suppliers' collaboration and reduce information asymmetry on costs. These ideas were discussed and screened in close interaction with the buying company personnel, thus identifying a shared set of actions/incentives to be activated to implement CT.

On suggestion of the buying company purchasing manager, the team noticed that it would have made no sense to define "one size fits all" incentives to manage different relationships with suppliers, underlining in this way the need to customize the right actions and incentives in relation to the specific "atmosphere". This conclusion was also confirmed by a pilot interview performed in an automotive lighting components company that already implemented a cost-sharing program with its customers, though mainly in the "one-way" direction. This interview provided us with a better understanding of the supplier's point of view on one-way cost sharing programs. Therefore, the research team realized it was important to capture the suppliers' point of view and to identify the conditions for the activation of actions/incentives included in the preliminary CT model.

3.5 Third Action Research Loop

Aiming to verify the validity of actions/incentives identified in the second research cycle, the team decided to interview a limited number of suppliers. Using the preliminary CT model, we were able to identify those suppliers willing and worth being engaged in CT initiatives. Operations managers and process improvement "champions" have been interviewed directly at the supplier company plant or through telephone interviews. We conducted 20 interviews in 15 supplier companies.

The analysis of data collected allowed us to refine the preliminary CT model. To validate the model, the team used it to select four suppliers with whom to collaborate to implement cost transparency and to identify an appropriate set of actions/incentives to achieve their commitment toward cost-information sharing.

4 Model Description

This research adopts to point of view of a buying company interested in sharing information on costs with some suppliers to eliminate wastes and generate value to be transferred to final customers. In fact, as argued by Lamming et al. (2005), though CT is a two-way process, with mutual benefits for the involved parties, its implementation may begin with the initiative of one of them.

At the beginning of CT implementation the key issue for the customer can be summarized as follows. Being S the whole set of buying company's suppliers, $S_1 \subset S$ is the subset including those suppliers which hold knowledge worth being shared with buying company to increase value delivered to final customer (also by reducing costs). $S_2 \subset S_1$ is the subset including those suppliers which are both worth and willing being involved in CT building projects with the buying company.

Therefore, implementing CT requires to address two main issues: (1) identify suppliers in S_2 subset using some variables to distinguish them in the whole supplier portfolio and (2) select what forms of incentives and actions to activate to practically implement CT with the selected suppliers, depending on each relationship "atmosphere".

As a first step in CT implementation programs customers should identify a limited set of suppliers whom to successfully cooperate in cost information exchanges. Literature and empirical evidences from on-field research suggest us that two main drivers may influence the initial choice: supplier willingness and buying company interest.

Supplier willingness has already been considered by Kulmala (2004) in his framework. This variable is important because companies consider cost information too important and secret to share it with their customers. This variable is influenced by two factors:

- *Customer bargaining power.* In many cases customers, to implement open books, lever on their own contractual power. This concept was confirmed by the manager of automotive lighting component company interviewed during the second loop of our action research. Major car assemblers have a considerable bargaining power as compared to most of their suppliers, thus they can force suppliers to share cost data. In Kulmala's framework this variable is labeled as "balance of power", as it considers the volume of business which is exchanged between the two parties.
- *Level of trust.* It is necessary a high degree of trust between the selected companies (Seal et al. 1999; Kulmala 2004) to start CT cooperation with suppliers.

Besides, we considered the buying company interest on the supplier, because it may be neither realistic nor efficacy to invest in CT programs with all the willing suppliers. This factor, not considered before in literature, has been highlighted by the action research process after the analysis of the existing supplier portfolio. In fact, initially we noticed that buying company managers were not interested to engage CT with several suppliers. We discussed with managers to understand what factors were able to influence their interest toward certain suppliers, and we identified the following:

- *Customer's advantages to start cost negotiations.* Sometimes information exchange can be "dangerous" for the buying company because it can determine frequent purchasing price corrections. This is particularly disadvantageous in rising prices conditions. For instance, some purchased items are characterized by a cost structure remarkably influenced by raw materials. As a consequence, a tight cooperation in information exchanges can facilitate suppliers in claiming price redefinitions when raw material costs increase
- *Customer's knowledge of the supplier production processes.* It is necessary to limit the surveying field to supply companies characterized by production processes not too complex or different from those of the buying company. This is a practical problem, because very complex or different processes do not allow

the buying company to provide significant contributions to improving upstream products or production systems, due to its limited competences and know-how

- *ABC classification—incidence of the purchases.* We introduced the Pareto's analysis factor because the buying company recognized the importance to start collaboration efforts from A-class suppliers, namely those with whom joint improvement programs can lead to significant cost reductions.

Using the categorization variables described above and the related subfactors, we employed a method similar to vendor rating/ranking to select those suppliers worth and willing being involved in CT programs. Each microfactor received a weighting calculated using the methodology proposed by Olsen and Ellram (1997). The supplier subset S_2 includes suppliers scoring high values on both the variables.

As a second step in CT implementation programs customers should identify a set of incentives and actions to activate in order to successfully cooperate with suppliers in cost information exchanges. Incentives and actions selection emerged from literature review and the brainstorming performed during action research. In order to identify what forms of incentives and actions to activate to practically implement CT, we focused on S_2 supplier subset. By interacting with buying company managers and analyzing the supplier portfolio, we identified two categorization variables that allowed to classify customer–supplier relationships in four groups. Each group requires specific forms of incentives and actions to stimulate suppliers to share cost information. The categorization variables are:

- *Type of sourcing strategy for each stock keeping unit* (SKU). The buying company adopts two main sourcing strategies, single sourcing, and multi-sourcing. Actions and incentives to develop cooperation based on CT should be selected depending on the specific sourcing strategy. In a single sourcing context the lack of competitor suppliers increases suppliers propension to share cost information. On the other hand, in multi-sourcing situations, suppliers able to provide the same level of quality and performance for the same SKU should be accurately identified. In fact, in this case it is important to prevent the buying company to use cost information to benchmark supplier in S_2 with suppliers delivering similar products with different (lower) quality level. This practice would preclude supplier trust and inclination to continue information exchanges.
- *Actual information sharing level with the supplier.* This aspect refers to the degree of existing information sharing at the customer–supplier interface (e.g., exchanges limited to design activities, exchanges encompassing also forecast data, etc.). We assume that increasing the degree of the existing information sharing, increases the availability to collaborate in CT projects. This variable is characterized by two aspects: supplier attitude toward information sharing and customer–supplier trust level. When the actual information sharing level is high, suppliers are more disposed to share detailed information also on costs. In particular, in our action research we noticed that suppliers involved in closer collaborations seemed more inclined to implement a detailed and complete cost information exchange. The less collaboration is close, the less suppliers are available to share detailed information. As suggested by Cooper and Slagmulder

(1999), we distinguish between two different levels of information sharing: partial and full cooperation.

Table 1 shows the four groups of customer–supplier relationships identified by crossing the two variables above. Each section of Table 1 describes the specific forms of actions and incentives to stimulate suppliers to share knowledge on costs. These actions have been identified in the brainstorming activity in the second action research loop and lately validated through the interviews to suppliers; as usually happens in the trial and error process of action research, some of these actions/incentives were discarded or redefined.

To effectively describe incentives and the actions to take to involve suppliers in cost information disclosure, we distinguish between *collaborative processes* (namely, the actions) and *incentives*. Collaborative processes can act as "carriers" of knowledge and information transfer between customer and supplier. They can be described as formalized means through which knowledge sharing occurs in a bidirectional way.

In particular, from the interviews we were able to identify two main "target" processes:

- *Co-design*. This collaborative process is exclusively concerned on design activities. Interviews allowed us to validate the effective availability of the suppliers to share cost data beginning from a collaboration in the design phase. Many suppliers underlined that in the preliminary phases of product concept definition is useful to start cost information exchange about products and processes to obtain meaningful results in terms of cost reduction. In this way it is possible to design new products especially tailored for the customer, avoiding to bear extracost to correct mistakes in the advanced design phases.
- *Co-makership*. This process is directly focused on collaboration in operations and logistics. By cooperating in these phases it is possible to update continuously cost information, thus allowing the buyer to monitor the supplier's conduct and carry out joint improvement plans.

On the other hand, incentives help to create the right conditions under which genuine CT and information exchange might be achieved. In particular, the following incentives were acknowledged by suppliers:

- increasing purchasing volumes by focusing on a restricted number of suppliers;
- sharing data on forecast and production plans for a better integrated planning;
- technical support even through plant investments, lean production plans, and value analysis in design activities;
- co-marketing programs provided by customers;
- support in negotiation with second tier suppliers;
- inter-organizational problem solving teams.

From Table 1 it is possible to match collaborative processes and incentives to each specific context.

As regards the co-design collaborative process, in contexts characterized by multi-sourcing strategy the buying company must prevent the supplier's to perception of opportunistic behavior and therefore protect each supplier's design know-how from competitor suppliers. In the context characterized by partial cooperation and single sourcing the limited information shared allows to develop co-design mainly in the early stages of the design process, because the customer has scarce knowledge to significantly contribute in cost reduction programs. On the contrary, in the context of full cooperation with a single supplier co-design can be extended in all stages of design to create a joint effort to cost reduction.

As regards co-makership processes, in contexts characterized by multi-sourcing strategy the horizontal knowledge sharing must be avoided to protect the suppliers' know-how. In contexts characterized by partial cooperation, due to the limited information shared by suppliers, co-makership collaboration can be limited to monitoring raw materials cost and to the identification of potential improvement areas, on which suppliers should focus their efforts.

Also incentives should be considered in relation to the specific context represented in the matrix. Basically, it is worth to notice that in contexts characterized by full cooperation suppliers are more inclined to react to a wider range of proposed incentives.

Especially in the single sourcing case, the buyer firm could leverage on inter-company problem solving teams both in co-design and co-makership processes, thanks to the close existing relationship with suppliers. Also technical support is suggested for instance by facilitating joint plant improvements or lean production initiatives supported by the buying company. Other possible incentives consist in sharing data on forecast and production plans, support in negotiation with second tier suppliers, increasing purchasing volumes and co-marketing.

Interestingly enough in the multi-sourcing context co-marketing is considered as a less appropriate or critical incentive. We define less appropriate or critical incentives in those actions whose implementation could be difficult or not always advisable. In fact, in multi-sourcing contexts the buyer can rely on co-marketing incentives only if he supports all competing suppliers involved in CT, without using preferential treatments and creating jealousies, thus avoiding the risk to affect the existing relationships. For the same reason, co-marketing is considered a less important incentive also in the multi-sourcing/partial cooperation context.

When analyzing relationships characterized by partial cooperation (right column in Table 1), due to the limited shared information by suppliers, the buyer can adopt mainly three incentives: negotiation with second tier suppliers and sharing data on forecast and production plans, and increasing buying volumes. In particular, single sourcing strategy relationships also facilitate co-marketing practices, while in multi-sourcing situations also increasing buying volumes are suggested.

In contexts of partial cooperation we found that limited information exchange and collaborative relationships complicate the implementation of inter-company problem solving teams due to expensive investments. Moreover, these inter-organizational teams could be useful only in the design phase, due to the limited improvements reachable in the following phases. Moreover, in this setting another critical incentive is the technical support. In the action research process this incentive especially in the form of plant investments supported by the buyer, seems not to be accepted by suppliers with partial cooperation, because they may fear to lose their independence.

5 Discussion of Results

The proposed model supporting CT implementation is an important outcome of the action research, as it allows to identify the variables affecting (1) supplier availability to start open book practices and (2) buying company interest to cooperate in cost information sharing with suppliers.

From the action research it emerged that, in order for CT program to be potentially successful, the buying company have to spend some time in identifying not only suppliers with whom to cooperate, but also purchasing items more suitable of being cost analyzed due to their simple operation processes. For instance, PVC or wooden kitchen shutters and all the wooden parts are particularly appropriate to be object of such cost analysis. Cost reduction actions may be more difficult with those purchased components categories where the quality and price levels are heavily determined since the early design phases. This is the case of kitchen hoods and other steel components as handles. Interestingly enough, the supplier screening phase is very selective. We started considering an initial set of 200 suppliers, and we identified 22 suppliers in S_2 subset (see Table 1).

As regards the second implementation phase of the proposed framework, each supplier reacted to the actions/incentives in a different way. In some cases, suppliers considered some incentives not worth enough to exchange detailed cost data, while at the same time others seemed more inclined. Although suppliers recognized the importance to start collaborations based on trust and information exchange, many suppliers underlined how their priority is focused on hedging and protect their know-how, even if risking of losing the supply contract.

To validate our model we selected four suppliers from S_2 supplier subset, one for each quadrant of the matrix in Table 1. In the case of the single sourcing and full cooperation context, the buying company obtained interesting initial results collaborating with a supplier of semi-finished components. This supplier demonstrated immediately the availability to start a cost information exchange, by accepting the promise of constant buying volumes, possible support in their negotiations with second tier suppliers and short/medium term forecasts exchange. The supplier shared cost information related to an existing product and the buyer considered this level of information obtained sufficiently detailed and complete. Therefore, the buyer can now monitor the supplier's processes and in case of variation of raw materials cost it will be possible to start collaboration plans aimed at cost reduction. Similar results have been found also in the case of multi-sourcing and full cooperation context.

Table 1 Incentives to enable cost transparency in the different customer–supplier relationships

Type of sourcing strategy for each SKU		Actual information sharing level with the supplier	
		Full cooperation	Partial cooperation
Single sourcing		*Collaborative processes* Co-design, in all the stages of design process Co-makership *Appropriate incentives and actions:* Inter-company problem solving teams both in co-design and co-makership Technical support Sharing data on forecast and production plans Support in negotiation with 2nd tier suppliers Co-marketing Buying volumes	*Collaborative processes* Co-design, above all in the early stages of design process Co-makership: limited *Appropriate incentives and actions:* Sharing data on forecast and production plans Support in negotiation with 2nd tier suppliers Co-marketing Buying volumes *Less appropriate incentives and actions:* Inter-company problem solving teams Technical support
Multi-sourcing		*Collaborative processes* Co-design, be careful on horizontal knowledge sharing Co-makership: be careful on horizontal knowledge sharing *Appropriate incentives and actions:* Inter-company problem solving teams both in co-design and co-makership Technical support Buying volumes Sharing data on forecast and production plans Support in negotiation with second tier suppliers *Less appropriate incentives and actions:* Co-marketing	*Collaborative processes* Co-design, be careful on horizontal knowledge sharing Co-makership: limited, be careful on horizontal knowledge sharing *Appropriate incentives and actions:* Buying volumes Sharing data on forecast and production plans Support in negotiation with 2nd tier suppliers *Less appropriate incentives and actions:* Inter-company problem solving teams Technical support Co-marketing

The buying company achieved significant positive results also in the multi-sourcing and partial cooperation context, collaborating with hood suppliers from the design phase and leveraging therefore on co-design processes.

Instead, in the single sourcing and partial cooperation context CT failed because of the sudden aggravation of the relationship with the supplier. The buyer tried to involve the supplier of shutters in cost exchange beginning from the design phase of a new product. Due to the decrease of raw materials cost, the buyer asked for lower prices, but the supplier preferred to refuse, leveraging on his bargaining power.

From the results discussed above, it is possible to get some practical research implications. The implementation of our framework made it possible for the buying company to start understanding CT program dynamics, managing to involve some of its suppliers in cost information disclosure. Owing to the investigation conducted with suppliers, this research demonstrated that the one-way use of cost information would preclude any cost data exchange, because these data would be used as a contractual weapon in the negotiation phase to exert pressure on supplier to reduce prices. Therefore, we found it was important to redesign the cost estimation software, in order to avoid pursuing supplier benchmarking and rather to use cost information to activate joint cost reduction programs by improving product or upstream processes. The buying company has also understood the necessity to respect suppliers' know-how and the importance to maintain the correct trust level to allow open books. These results are direct outcomes of the action research process conducted in close relationship with the buyer firm and its suppliers: researchers played an important role advising the buying firm after the careful literature review and the direct interaction with suppliers.

6 Conclusions

This chapter proposes a framework to support buying companies in the selection of customer–supplier relationships worth being developed into CT, and in the identification, for each of these relationships, of appropriate forms of actions/incentives to stimulate suppliers to share cost information.

The main contribution of this research lies in its attempt to fill the gap in the literature regarding the lack of guidelines and structured methods to effectively implement CT. In particular, our model provides an original approach to classify customer–supplier relationships with the aim of identifying those suppliers that could successfully be involved in two-way CT programs. Our framework integrates Kulmala's one (Kulmala 2004), because it explicitly includes and operationalizes a new classification variable the action research identified as pivotal when choosing suppliers with whom to cooperate in CT programs: the buyer's interest. In addition, unlike existing models on CT implementation, our framework is able to provide managers with a set of actions/incentives to engage suppliers in CT practices. Interestingly enough, the framework does not suggest "one size fits all" solutions, because the action research provided evidence that different customer–supplier relationship "atmospheres"—in terms of actual cooperation level and type of sourcing strategy—require different incentives. Therefore, our framework can be a useful support to customize the set of actions and incentives to be launched to stimulate information exchanges on costs during CT implementation.

This research, like most studies suffers from limitations. The most obvious is that a single action research, while conducted on a sample of 19 customer-supplier dyads, can be hardly generalized to the overall population of manufacturing entities. Large sample data collection efforts will be needed to test the proposed findings and provide sufficient support to make our prescriptions more robust. Additionally, the proposed model could be tested in other industries to investigate whether our findings are still valid outside the modular kitchen industry and what kind of adjustments are needed.

Further research could develop our model from an agency theory perspective in order to understand how incentives can be used to cope with information asymmetry which influences trust between parties, therefore determining the success/failure of any CT initiative. We think that Manatsa and McLaren's recent study (Manatsa and McLaren 2008) can be a good starting point to address such research hint.

Moreover, we argue that our model can be improved using a "dynamic perspective", namely determining what actions to engage in CT programs those suppliers reluctant to cost data disclosure. This means to move them from low-left cells in the selection matrix (Table 1) to the subset S_2 of suppliers not only worth, but also willing being involved in CT programs.

References

H. Agndal, U. Nilsson, Supply chain decision-making supported by an open books policy. Int. J. Prod. Econ. **116**, 154–167 (2008)

H. Chin-Chun, V.R. Kannan, T. Keah-Choon, G. Keong Leong, Information sharing, buyer-supplier relationships, and firm performance—a multi-region analysis. Int. J. Phys. Distrib. Logist. Manag. **38**(4), 296–310 (2008)

R. Cooper, R. Slagmulder, *Supply chain development for the lean enterprise—interorganizational cost management* (Productivity Press, Portland, 1999)

P. Coughlan, D. Coghlan, Action research for operations management. Int. J. Oper. Prod. Manag. **22**(2), 220–240 (2002)

H. Håkansson, *International marketing and purchasing of industrial goods* (Wiley, Chichester, 1982)

P. Kajüter, H.I. Kulmala, Open-book accounting in networks. Potential achievements and reasons for failures. Manag. Account. Res. **16**, 179–204 (2005)

H.I. Kulmala, Developing cost management in customer–supplier relationships: three case studies. J. Purch. Supply Manag. **10**, 65–77 (2004)

H.I. Kulmala, J. Paranko, E. Uusi-Rauva, The role of cost management in network relationships. Int. J. Prod. Econ. **79**, 33–43 (2002)

R.C. Lamming, *Beyond partnership: strategies for innovation and lean supply* (Prentice-Hall, London, 1993)

R.C. Lamming, Squaring lean supply with supply chain management. Int. J. Oper. Prod. Manag. **16**(2), 183–196 (1996)

R. Lamming, N. Caldwell, W. Phillips, D. Harrison, Sharing sensitive information in supply relationships: the flaws in one-way open-book negotiation and the need for transparency. Eur. Manag. J. **23**(5), 554–563 (2005)

P.R. Manatsa, T.S. McLaren, Information sharing in a supply chain: using agency theory to guide the design of incentives, *Supply Chain Forum: An International Journal*, Vol. 9, No. 1 (2008)

R. McIvor, Lean supply: the design and cost reduction dimensions. Europ. J. Purch. Supply Manag. **7**, 227–242 (2001)

J. McKay, P. Marshall, The dual imperatives of action research. Inf. Technol. People **14**(1), 46–59 (2001)

J. Mouritsen, A. Hansen, C.Ø. Hansen, Inter-organizational controls and organizational competencies: episodes around target cost management/functional analysis and open book accounting. Manag. Account. Res. **12**, 221–244 (2001)

D. Näslund, Logistics needs qualitative research—especially action research. Int. J. Phys. Distrib. Logist. Manag. **32**(5), 321–338 (2002)

R.F. Olsen, L.M. Ellram, A portfolio approach to supplier relationships. Ind. Mark. Manag. **26**, 101–113 (1997)

J.O. Piontkowski, A. Hoffjan, Less is sometimes more: The role of information quantity and specific assets in the propensity to engage in cost data exchange process. J. Purch. Supply Manag. **15**(2), 71–78 (2008)

W. Seal, J. Cullen, A. Dunlop, T. Berry, M. Ahmed, Enacting a European supply chain: A case study on the role of management accounting. Manag. Account. Res. **10**, 303–322 (1999)

R. Westbrook, Action research: A new paradigm for research in production and operations management. Int. J. Oper. Prod. Manag. **15**(12), 6–20 (1995)

R. Yin, *Case study research: Design and methods*, 3rd edn. (Sage Publications, Inc., Thousand Oaks, 2003)

Chapter 9
Learning on the Shop Floor: The Behavioural Roots of Organisational Knowledge

John D. Hanson

Abstract If it is accepted that knowledge is a valuable asset of a firm, then it follows that learning or the ability to learn is equally important. This study used shop-floor observations to understand how this learning takes place in an organisational setting. It was discovered that for improvements in organisational knowledge to be realised, two distinctly different components had to be present. The first component consists of the performative aspects of the task at hand: techniques, resources, scientific facts, and the like. These elements are readily codified, stored, retrieved, and copied. They are cumulative in the sense that acquiring a new technique does not eliminate the ability to use an existing appropriate one . The second component is the underlying logic by which techniques are selected and applied to achieve a goal. This component corresponds closely to Plato's description of knowledge as justified belief. This component is not readily observable and is not cumulative in the sense that one cannot simultaneously hold conflicting beliefs. Because the context and justification of belief are defined by cultural norms, this introduces a strong behavioural component to organisational knowledge. The existence and importance of this largely invisible component have significant implications for managers who wish to promote organisational learning and for researchers who wish to study it.

J. D. Hanson (✉)
School of Business Administration, University of San Diego, 5998 Alcalá Park, San Diego, CA 92110, USA
e-mail: hansonj@sandiego.edu

I. Giannoccaro (ed.), *Behavioral Issues in Operations Management*,
DOI: 10.1007/978-1-4471-4878-4_9, © Springer-Verlag London 2013

1 Introduction

Few would challenge the premise that knowledge is an important asset of a firm. There is also a corresponding belief that this importance to the firm and to the economy at large is increasing with time, attributable primarily to the increasing technical complexity of our world (Drucker 1993; Bettis and Hitt 1995; Mukherjee et al. 1998). That being the case, it must also be true that learning (in the general sense of improving the state of knowledge) is a process critical to firm success.

The recognition of knowledge as an important *resource* of the firm is not new (it can be seen clearly in Penrose 1959 for example), but increased attention to this fact has coincided with the emergence of the *Resource-Based View* of the firm (RBV), a term introduced by Wernerfelt (1984), in the field of strategy. Arguing the economic origins of the RBV, Williamson (1981) makes the case that knowledge is a source of economic rent. Extending this reasoning, Grant (1996) proposed the Knowledge-Based View (KBV) of the firm. Central to this view is the argument that knowledge is actually a firm's only source of economic rent on the basis that it is the only resource whose use incurs no opportunity cost. Many researchers have made statements to this effect (Prahalad and Hamel 1990; Barney 1991; Nelson 1991; Leonard-Barton 1992; Cyert et al. 1993; Henderson and Cockburn 1994; Nonaka 1994; Bates and Flynn 1995; Nonaka and Takeuchi 1995; Kogut and Zander 1996; Miller 1996; Spender 1996; Davenport and Prusak 1998; Nahapiet and Ghoshal 1998).

One result of this level of attention is a vast literature on organisational knowledge in which the subject is approached from a variety of viewpoints with an equally wide range of findings. To clarify the discussion, this article will adopt a narrow definition of the term organisational knowledge that addresses both parts of the term: what is knowledge, and what is it that makes it organisational in scope (as opposed to common or individual knowledge)?

A key issue in knowledge is the *appropriability* of the rents that derive from it (Collis and Montgomery 1998). In that context, it is an important distinction whether knowledge is uniquely a property of the organisation (which would then capture the rents), or a property of the individuals who happen to make up the organisation at the time (and who should therefore be able capture all or most of those rents through above-normal salaries). In the work that follows, the term *organisational knowledge* will be used to mean a form of knowledge that is uniquely a property of the firm; something that the firm can claim as its own and is robust against disturbances such as employee turnover.

The question of what constitutes knowledge is more contentious. Nonaka and Takeuchi (1995) note that modern Western philosophy has been dominated by the *Cartesian Split*, the paradigm of Descartes in which knowledge, and by extension truth, are independent absolutes which can be discovered and are therefore independent of the person. Polanyi (1958) however concluded that this view is ultimately not sustainable and that knowledge is, in the final analysis, something unique to the knower. In this he falls back on a definition of knowledge as *justified belief,* a definition that Nonaka and Takeuchi (1995) attribute to Plato. It is the role of belief and the context of justification that makes this an individual or personal form of knowledge.

However, our interest is primarily in knowledge as a property of an organisation, and it is not clear that definitions centred on individuals are relevant when applied to organisations. Cyert and March (1963) first proposed that an organisation could learn independently from its individual members but did not leave us with a comprehensive definition of organisational knowledge. Grant (1996), in proposing the Knowledge-Based View of the firm, avoided the issue by ascribing knowledge purely to the individuals comprising the organisation. Teece et al. (1997) also circumvented the issue by arguing that what the organisation possesses is the complementarity of the knowledge of individuals assembled within the organisation. Nonetheless, the term *Organisational Knowledge* is very much part of our vocabulary and many feel that it is a distinct form of knowledge. Specifically, Hedberg (1981) describes it as being a "robust property" of an organisation.

To arrive at a useful definition of organisational knowledge, we follow the lead of Pentland (1992), who drew on the philosophy of American pragmatism, and specifically the work of George Herbert Mead, William James and John Dewey. This school also rejects the Cartesian view of knowledge and it is from their writings (specifically: Dewey 1916; Dewey and Bentley 1949) that we derive the idea that knowledge is *situated,* that is to say: defined only relative to a situation. Accordingly, we adopt the definition of organisational knowledge as *situated performance,* which means the ability to deliver a satisfactory performance in a defined set of circumstances. This can be justified by the fact that organisations generally have a defined purpose that defines the situation.

Many similar definitions or statements about organisational knowledge have been offered. Amin and Cohendet (2000) speak of the firm as: *a social institution, the main characteristic of which is to know well how to do certain things* (p. 96). In the same vein, Orlikowski (2002) equates knowledge with *effective action* as do Sabherwal and Becerra-Fernandez (2003).

With knowledge defined as a form of capability, there is still the issue of how it becomes a property of the firm; where it resides, as it were. Here we turn our attention to the subject of *routines.* We do so because many researchers have concluded that routines are the means by which knowledge is embedded in an organisation (Levitt and March 1988; Cohen 1991; Amundson 1998; Orlikowski 2002). While no single researcher can claim priority on the concept of routines as organisational knowledge, the idea is clearly present in the work of March and Simon (1958) who observed that what we are calling organisational knowledge existed in the form of *performance programs,* which we now more commonly refer to as *routines.* These routines are fundamental to production and operations management. Although they drew our attention to routines, March and Simon (1958) did not describe how they were encoded in organisations or how they were created or changed.

Progress has been made on the anatomy of routines by Feldman and Pentland (2003) who expanded the basic concept of routines to include the potential for variation and selection, and therefore evolutionary, incremental change. More recently, Feldman and Pentland (2005) and Pentland and Feldman (2005) have argued that to be understood, routines must be considered as consisting of two layers; the *ostensive* and the *performative* (plus associated *artefacts*). The performative layer is the observable sequence of actions that constitute the routine as it is practiced in the moment. To an outside observer, it might seem that this would be a complete description of the routine. The ostensive part is the relatively unobservable underlying purpose or intent of the routine. The need to understand the ostensive layer was motivated by their observation that the performative layer tended to vary unpredictably (to our outside observer at least) in response to changing conditions. If the routine is the embodiment of organisational knowledge, the ostensive part is at least as important a determinant of situated performance as the performative part. It is the feedback loop by which the performance is assessed relative to the situation and by which the process is adjusted to suit.

By extension, if performative and ostensive components are both essential components of organisational knowledge, then beyond some point, organisational learning must involve changes to both parts. This is precisely the point made by Argyris and Schön (1978) when they coined the term *"double loop learning"* to describe the case where the ostensive part is examined and reconsidered. They did not use the term ostensive; instead they spoke of the "theory in use" that provided the belief that a certain set of actions would result in the desired outcome.

In Operations Management, the literature has focused heavily on performative aspects ("tools") of routines, with a resulting lack of explanatory power. To cite a well-regarded paper as an example, Flynn et al. (1995) analyse the performative aspects of organisations that can be discerned from survey data, and obtain results that hint strongly at the importance of an (unidentified) underlying ostensive aspect. Using different terminology, Miller (1996) showed recognition of this gap by stating that competitive advantage lies: "not in specific resources or skills, but in orchestrating themes" (p. 509).

With organisational knowledge defined as situated performance we can derive an operational definition of *organisational learning* as: *an increase in situated performance*. We should probably focus on deliberate rather than accidental improvements in performance, so this definition can be supplemented by one proposed by Argyris and Schön (1978): *the detection and correction of error*. Equivalence of the definitions can be established by regarding "error" to be any unsatisfactory level of situated performance. Significant in this definition, and very much to the authors' central point is that the assessment of "error" depends not only on the level of performance, but also on the perception of what it should be and how it should be measured.

Whether this action of learning consists of creating new knowledge or applying existing knowledge may be an artificial distinction (Starbuck 1992) but what we can know is that the end result is an improved capability on the part of the firm. However, while the knowledge created may be a property of the firm, the actions taken are by individuals who must be presumed to respond to behavioural influences and limitations. If we are to understand the mechanism by which knowledge gets into practice, we need to understand more about why people do what they do.

2 The Shop-Floor Perspective

Researchers are generally obliged to study firms from the outside. As a result, it is not surprising that much of what we know about organisational learning is expressed in terms that describe the behavior of the firm as a monolithic entity. An important example of this is the term *Absorptive Capacity* (Cohen and Levinthal 1990). This is a measure of a firm's ability (or perhaps: propensity) to learn, and while it may provide a useful way to categorise and describe firms, it tells us relatively little about how this takes place. When a firm is seen as a single entity, the analogies to an individual person are inviting and Cohen (1991) and Cohen and Bacdayan (1994) make the case that, just as individuals store their learning in either procedural or declarative memory, so too do organisations have distinctly different modes of retaining their knowledge. Organisational routines are argued to be the analogue of procedural memory—the largely unconscious "how to" that functions relatively independent of logic or data. Again, while this provides useful labels, the terms tell us little about how the knowledge got there in the first place or how it is changed.

When we attempt to look inside the organisation, typically through the use of survey instruments or structured interviews, we encounter an additional difficulty in that such data is almost always collected at a single level within an organisation. In a recent study on innovation, Rothaermel and Hess (2007) noted that the antecedents to innovation took place at the individual, firm, and network levels. To focus on only one level, they noted, could lead to two serious problems, or as they put it:

> First, concentrating on only one level of analysis implicitly assumes that most of the heterogeneity is located at the chosen level, whereas alternative levels of analysis are considered to be more or less homogeneous. Studies of firm-level heterogeneity assume, for example, that significant variation occurs at the firm level of analysis, whereas individuals are more or less homogeneous or randomly distributed across firms. Second, when focusing on one level of analysis, researchers implicitly assume that the focal level of analysis is more or less independent from interactions from other lower- or higher-order levels of analysis. Firm-level heterogeneity, for example, is assumed to be relatively independent from individual- or network-level effects. *Taken together, the assumptions of homogeneity in, and independence from, alternate levels of analysis are serious concerns that could lead to spurious empirical findings* (p. 899, italics in the original).

Given these limitations on existing research, it becomes apparent that to gain additional insight into the process of organisational learning it will be necessary to adopt different approaches. In this we have some guidance from Ohno (1988) who was adamant that learning takes place where the action is, hence *genchi genbutsu* (roughly translated as: "get out on the shop floor and see what is really going on," (Dennis 2002, p. 141).

Doing so introduces an ethnographic component to the research, in which the output is *cultural interpretation* as opposed to simple data (Hoey 2011). The nature and goal of such research is to gain enhanced insight, at the expense of wider generalizability. The introduction of a cultural dimension raises issues of methodology and interpretation, summarised by the *etic/emic* divide (Martin 2002, p. 36). The distinction in this divide is that the researcher's point of view is either within the culture, or outside of it. Each point of view will fail to see things that the other will. Most typically, the etic (outsider) view will fail to see the significance attached to various symbols or actions, while the emic (insider) view will fail to see that certain taken-for-granted truths are not universal. In addressing the shortcomings of the etic view, Bartel and Garud (2003) note that: "to see and understand narratives, the researcher must become *semi-native*" (p. 337, emphasis in the original). This argues for the selection of a participant-observer model of observation (Martin 2002, pp. 48, 210) which can bridge the "insider" and "outsider" views of an organisation.

3 The Study

The study described here was motivated by the objective of shedding light on the process of organisational learning and sought to do so observing the process in action. This required a suitable setting and an appropriate method. As discussed above, the participant-observer method was felt to be appropriate for this question because the goal was not simply to record actions or results, but to understand the logic behind the actions, as interpreted by the subjects themselves.

A setting had to be chosen in which this model could be employed effectively. For efficiency, it was necessary to select settings in which organisational learning was expected to occur. For validity it was necessary to narrow those down to ones that would be relatively free of confounding influences. An additional level of validity is established by repetition: there are numerous examples of in-depth research of this kind that result in single-sample anecdotes, whose relevance to each other is hard to establish. Ideally what was wanted was a group of settings that conformed to some consistent standards so that we could compare the actions of different (but comparable) people in different (but comparable) organisations addressing different (but comparable) problems.

The class of activities known as Kaizen Events, as described by Melnyk et al. (1998), was selected as satisfying these requirements. These events have useful characteristics for the purposes of this study. By their very nature they are designed to promote and capture organisational learning. They are also bounded in their scope, their duration and their slate of participants. This bounding removes many unobservable factors that could otherwise confound the observations. A key choice was whether to seek out Kaizen events from different corporations or to study events taking place under a single corporate umbrella. The latter alternative was eventually chosen to take advantage of the greater uniformity of approach. The common mode problem was minimised as much as possible by insisting that the events studied should take place in distinctly different operating units of the corporation.

A target corporation (hereafter: the Corporation) that was known to conduct significant numbers of Kaizen events was approached and asked for permission to observe some of the events. The Corporation, through its individual operating companies, is engaged in the manufacturing of a wide range of products used in commercial and residential construction and remodelling. At the time of the study, there were over 50 of these autonomous operating companies, organised into five product-based business lines.

It was agreed that a researcher would be allowed to attend a number of events acting as a participant-observer. Within the Corporation, there was a range of events to choose from, but many were not totally suitable for the study. Only events focused on process improvement in manufacturing operations were considered, and any event that was conducted in two parts or that represented a continuation of an earlier event was ruled out. This two-part structure was fairly widely used in the Corporation, and while it has certain practical advantages, it compromised the ideal of a bounded event because the actions of the participants could not be observed in the intervening periods. As a result of this selection process, three research cases were identified. The Kaizen events comprising these cases took place in a six-month time window, during which there were no major shifts in the economic environment experienced by these companies. The key features of these three events are summarised in Table 1.

There are obvious limitations to this approach. Organisational learning is only being observed in one particular type of setting and it would be improper to assert that the observed mechanisms would broadly applicable to other situations. The objective however, was insight, not generalizability and method of data collection and analysis was designed to support this. The collection and presentation of data presents some difficulties in this type of research because the approach is *inductive*. By that, it is meant that we observe evidence (outcomes) that would verify potential hypotheses in order to arrive at more general propositions.

The issue is with the method by which observations (representing potential hypotheses) are distilled into propositions. Hunt (1991) offers two choices, neither of them entirely satisfactory. If the observations are quantifiable, and enough data can be collected, then propositions can be derived statistically, as in exploratory factor analysis. This is a somewhat weak method, since propositions derived in this way cannot be falsified; one can only say that they are supported or not within certain confidence limits.

When observations are not quantifiable or are few in number, both of which are the case in this research, the remaining alternative is pattern matching, which is the search for distinctive and recurring patterns in the data (Hunt 1991). The realities of the social sciences dictate frequent use of the method, and it is advocated in this context by Kaplan (1964). While this method can generate useful insight,

Table 1 Summary of events

Description of process	Central goal of event	Team composition	Summary of major results	Comments
Event #1—Plumbing valve body fabrication, automated brazing carousel	Reduction of setup time to change fixtures for part # changes	1. Internal facilitator (manufacturing engineer) 2. Outside facilitator 3. Tooling engineer 4. Assembly worker 5. Lead hand/setup specialist 6. Warehouse supervisor 7. Tool and die maker 8. Quality control rep. 9. Researcher	Labour content to change fixtures reduced by ~50 %. Numerous other small improvements, most significant: improved post-cooling to reduce sticking on fixtures, improved airflow (cooling) to reduce breakage of brazing wire during automatic wire feeding	Not all opportunities to reduce machine downtime were implemented. Time reduction was not validated in practice but is believed to be accurate. No scheduling changes were implemented at this time
Event #2—Metal fabrication facility; cutting, bending, machining, welding, painting of machine frames	Reduce work-in-process inventory between welding and painting operations, also reduce floor space	1. Director of operations (internal facilitator) 2. Outside facilitator 3. Quality control rep. 4. Operator: assembly area 5. Operator: cable area 6. Operator from paint line 7. Researcher	Developed revised painting schedule and "supermarket" rack between operations to reduce WIP by 75 % by cycling through paint colours four times/shift instead of once. Minor other housekeeping improvements	WIP and floor space reduction dependent on scheduling, not necessarily sustainable. Note that director title may mislead; manufacturing engineer would have been more typical designation

(continued)

Table 1 (continued)

Description of process	Central goal of event	Team composition	Summary of major results	Comments
Event #3—Final assembly and packing of shower control valves (retail and contractor packs)	Replacement of carton packing bench to be more ergonomically sound. Rearrangement of component storage to facilitate SKU changes	1. Internal co-facilitator from human resources 2. Internal co-facilitator; production supervisor 3. Production team leader for area, first shift 4. Production team leader for area, second shift 5. Chem lab tech., 3rd shift 6. Test technician 7. Assembly operator 8. Researcher	Revised bench designed, built and implemented. Stock of components rearranged by family to reduce handling and distance travelled during changeovers. Expected to improve housekeeping, reduce errors and damage (not quantified). Mistake-proofing implemented to prevent recurrence of a known issue	Key difference from other events is lack of quantitative targets—seemed to reflect higher level of confidence that team would do the right thing

Hunt (1991) points out that it suffers from a lack of intersubjective confirmability. By this he means that if two researchers extract different patterns from the same data, there is no test to determine which is better. Under these circumstances, the best that one can hope for is a degree of consensus. In this research this issue was addressed through the use of a panel of experts who were asked to comment on the validity of some of the key patterns observed.

Since validity is of critical importance in case study research, particularly internal validity (Anderson et al. 1999) additional steps were taken. Purposeful case selection has already been described, as has *validation* by the expert panel. As recommended by Yin (1994), *triangulation* was employed in the form of the use of multiple sources of evidence to address the research question. This was possible in these cases because, in addition to the observations of activities, the full range of data and documents used by the project teams was made available to the researcher. To guide the actual collection of observations a ten-item *research protocol* (Ellram 1996), was developed from a comprehensive review of the literature on organisational learning. This was used as an event checklist to ensure that each case was addressed systematically in search of evidence supporting or contradicting propositions derivable from the literature. The actual transcripts taken during the events were augmented with additional details filled in at the end of each work day. Copies of these notes were returned to the team leaders after the event for *verification* and correction of any errors of fact. Once approved, axial or pattern coding was used (Miles and Huberman 1994). That is, each item of each transcript was reviewed to determine whether it constituted evidence for the presence or absence of any of the protocol elements. These observations were then sorted to provide a summary of the relevance of the protocol element in explaining what had taken place.

4 Key Findings: The Nature of Organisational Learning

As described above, each of the case settings involved a conscious effort to effect process improvements which, if successful, would be evidence that organisational learning had taken place (in the sense that *situated performance* would have been improved). Across the three settings, 38 identifiable process improvements were implemented, but our interest is not in the improvements themselves, rather the question is how they originated and came to be implemented.

It was initially supposed that this would involve examples of knowledge transfer, given that the central theme was the implementation of Lean techniques that are well-documented. In that view, organisational learning would consist of copying knowledge from one "bin" to another (Walsh 1995) or converting from one type to another (Nonaka 1994). As a result, part of the data collection was a tabulation of where people turned for inspiration when faced with a problem.

These results are summarised in Table 2, but this is incomplete because in every instance the first place everyone turned was to their own experience to see if

Table 2 Source of information

Case #	Problem-solving examples	Consult with experts	Observe existing examples	Look up documentary records
1	17	4	3	1
2	14	3	1	1
3	7	0	1	0
Total	38	7	5	2

Experts are individuals outside the project team with experience relevant to the problem at hand

Existing examples are previously—implemented problem solutions elsewhere in the company

Documentary records are company records specific to the process or product (as opposed to
   supplier catalogs, for example)

they had seen something like this before that could be applied in the present situation. Since this was universal, it is not listed in the table. The second choice was to tap into the experience of others by seeking out experts with relevant experience. The key point in this table is that these first two categories (own experience and experiences of others) are *efficient* in the sense that they contain not only specific details, but context and interpretation as well. The last two categories are *inefficient* because they lack these things. In fact, documentary sources, even when apparently directly relevant to the problem at hand, proved to be generally unusable unless an expert could be found who could explain what was done, why, and the story behind the story. As a result, use of these resources was avoided even when they were available. As a simple example of this, in the first case a question arose as to whether a certain modification to a gas line would be allowed by the code. Now the text of the code would have provided a definitive answer, but the response was simply to call up a plumbing contractor that the company used regularly and ask him the question. He was able to answer this immediately, and more importantly he was also able to answer the real question which was whether or not an inspector would pass it, thereby addressing issues of context and interpretation.

The use of existing examples is an interesting situation, and an example will serve to illustrate the mechanism involved. In the third case, one of the tasks was to replace a packing bench with a new one that would be ergonomically friendlier. (The old one required operators to walk around it repeatedly and to reach excessive distances). It was observed that a bench on another assembly line in the plant had a good reputation in this regard, and two team members were dispatched to study it with the goal of replicating its key features. Since they could not copy it exactly, what they had to do was essentially to reverse-engineer it to discover the design rules for creating an ergonomically friendly work station. This they did in a very tacit way, taking only a few measurements and writing nothing down. Whether they discovered all of the logic embedded in the original design is unknowable and it is reasonable to assume that they also had a few preconceived ideas of their own about how such a bench should be built. When they were finished, the new bench was an improvement, so we can say that learning took place. However, it would not be fair to call this a transfer of knowledge; more realistically we would have to describe it as a creation of new knowledge, albeit heavily influenced by existing example.

This example serves to illustrate the central theme of the findings: that knowledge is not transferred or converted; it is created anew in a form unique to its situation and from a combination of factors. This is fully consistent with the definition of organisational knowledge as situated performance, and is also analogous to the concept of situated cognition (Rosch and Lloyd 1978). Furthermore, each instance of organisational knowledge consists of two distinct parts. This insight is also reflected in the literature, but most explicitly by Pentland and Feldman (2005). In describing routines as the unit of analysis within which organisational knowledge is embedded, they characterised routines as being comprised of tangible, observable elements such as work practices and procedures (in this case, the bench itself), and an unobservable component they called the ostensive element (the design rules, and if necessary, the operators' understanding of how to use it). To be more precise, they characterised routines as containing three parts, the third being the physical artefacts such as fixtures or line layouts. For our purposes these serve much the same purpose as the performative elements such as work instructions. The ostensive component is hard to define precisely, but corresponds to the conceptualisation of the problem to be solved and the appropriate range of approaches for doing so. What was found, as illustrated in the above example, is that both parts had distinct impacts on situated performance. We can state this more broadly as a proposition:

P1: Organisational knowledge (and by extension, organisational learning) must be understood as consisting of two distinct components, the performative and the ostensive.

Note that this is similar to the often stated distinction between knowing "what" and knowing "how", that Edmondson et al. (2003) also characterise as a distinction between codified and tacit knowledge. In terms of their impact on performance, the terms performative and ostensive are preferred here because they are more descriptive of what is actually going on.

An example was presented above showing how both elements were essential to the creation of an effective work station. Armed with this insight, every instance of changes to routines (learning) could be analyzed the same way and in many cases the linkage between ostensive and performative elements was obvious. It is the exceptions that are interesting however, and the following detailed example will illustrate how a mismatch between the components impacts organisational learning and the capacity for such learning.

In the first of the three case studies, a primary goal was to reduce the time taken to change over an automated brazing carousel from one product to another as part of the company's progress to Lean Production. In the current state of affairs a changeover took from 1-1/2 to 2 h and it was felt that this amount of downtime could not be afforded during regular shifts, so production was batched and changeovers were normally conducted on third shift when no production was scheduled. With the goal of setup time reduction in mind, the training and orientation portion of the Kaizen event featured selected training materials on the subject of quick changes so that the participants were exposed to a wide variety of techniques that they might be able to apply.

One of the lead hands who routinely performed the changeovers was a member of the team, which was significant because he would have to implement the subsequent revisions. What happened was that as long as the efforts were directed towards simply reducing the amount of work required for changeovers, there was little debate and the proposed improvements were readily accepted and implemented. A great deal of creativity was demonstrated, and some of the changes were quite ingenious.

However, it is also possible to reduce the actual downtime by other means such as pre-staging some of the work before the shutdown or deferring some of the clean-up until after it. It is also possible to add workers while the machine is stopped even though they may not be performing as efficiently as they could if they weren't getting in each other's way (Hall 1983). When such suggestions created the possibility of further decreasing machine downtime at the expense of adding man-hours, the reaction was interesting. These proposals were not challenged; they were simply ignored. In this event it was proposed and eventually (somewhat reluctantly) agreed to build some pre-staging racks for setting up the new fixtures in advance, but very little progress was made on implementing this change. It was observed that, somehow, those working on it seemed to keep finding other tasks to have higher priority. When asked about this, and the idea of bringing in a second person to speed the changeover, the individuals most directly involved didn't really want to challenge the training materials, but made it clear that these ideas didn't make sense to them in their situation.

Upon further questioning, the participants demonstrated that they were motivated primarily by an interest in minimising the hours booked for fixture changes. This of course is not the point of setup reduction under Lean Production—these workers were solving a different problem. Although apparent progress was made on the performative elements of the problem, it was the hidden ostensive component that ultimately limited the organisation's ability to "learn" in the sense of improving situated performance. This is the point made by Barnard (1938) who wrote that: "An intelligent person will deny the authority of that ... which contradicts the purpose of the effort as *he* understands it" (p. 166, emphasis in the original).

Although the reported results from this event were good (50 % reduction in setup times), it represented a failure of sorts in the organisational learning process. What is interesting about this failure is that it occurred in a setting that should have been highly conducive to learning: training was provided, resources were made available as needed and there was visible evidence of upper management support; all things commonly cited as success factors. The team members had all the tools they needed and they had an adequate understanding of the principles of Lean, yet they had not adopted a truly Lean perspective; instead they were achieving what Emiliani (2007) refers to as "False Lean".

In examining the reasons for this, it became apparent that knowledge of performative aspects of business processes is cumulative. That is to say, an organisation can be more or less knowledgeable on a continuum and it is reasonable to accept that the training referred to above had made the organisation more knowledgeable with respect to the performative aspects of Lean. The same, however, is not true of the ostensive elements. While a person or organisation may simultaneously know how to perform multiple tasks, they cannot simultaneously hold different opinions about which is the best under the circumstances. In that sense, the ostensive component of organisational knowledge is something of an either/or proposition. Although training had been provided in the ostensive aspects of Lean Production, and the participants understood it well enough, it had failed to displace the existing mind-set of those charged with implementation. We can state this as our second broad proposition:

P2: If organisational learning requires changes to a goal or the path to achieving that goal, then displacement of the previous goal or approach is an essential element of the learning.

It is clear that when we talk about a goal or an approach to a goal, we are discussing the ostensive component of organisational knowledge. Goals and approaches are somewhat interchangeable depending on the level we are looking at within the organisation. For example, a CEO may decide that to achieve a goal of increased shareholder value, it is better to switch from a strategy of rigorous cost-cutting to one of radical product innovation. This represents simply a change in approach and while there may be significant operational challenges, it is not all that difficult conceptually. At the shop floor however, this is a fundamental change of goal and requires a significant reconceptualisation of the purpose of the enterprise. It is not surprising that change initiatives often stall when they reach this level.

The concept of displacement corresponds well to the term *unlearning*, the importance of which has been noted by Weick (1979), Schein (1993) and Pentland (1995). Recognising the need for displacement is much easier than prescribing how to achieve it. What does seem apparent is that failure of the present logic is not sufficient since individuals will remain committed to their plans of action in the face of substantial evidence of their lack of utility (Mitroff and Mason 1974). Some simple examples from our cases illustrated this. There were some problems that had to be dealt with along the way (two examples: parts sticking to fixtures, brazing strips breaking at the feeders) where the proposed solutions depended on beliefs about the root causes. In general, failure of the proposed solutions did not cause the operators to re-assess their beliefs about root causes. That only occurred when they were faced with clear evidence that an approach based on a different causal mechanism did work.

Unfortunately, for more important issues where the payoff is in the future and not necessarily deterministic (for example, success of a market strategy may depend on a competitor's response), clear evidence is hard to come by. In these situations we see that the ostensive component of organisational knowledge corresponds well to Plato's description of knowledge as "justified belief" (Nonaka and Takeuchi 1995).

When organisational learning requires change to the ostensive component (the belief), it is not only necessary that the new approach be justified, it must be better justified than what it is to replace. This forces us to confront the basis on which existing beliefs are justified and is the reason that this article positions organisational learning as a fundamentally behavioural issue.

These cases provided no very good examples of how this displacement process might actually work, in spite of the fact that efforts were made during training to instil a new "justified belief". All of these events were focused on making progress towards Lean Production, and to that end, each event included training about the philosophy and benefits of Lean. Most participants seemed to understand this material and could articulate it, but subsequent conversations made it very clear that this training had done nothing to change their prevailing mind-sets. To capture this in an abbreviated way: in all three companies, it seemed that floor-level workers had an intuitive understanding that their mission was to maximise output per unit of their time. Nothing in the training had caused them to reject this view in favour of other approaches, as for example when a Lean production schedule with smaller lot sizes would require them to work harder and actually produce less. This clearly did not make sense in their world. To some extent that serves as evidence for the difficulty involved when there is a lack of clear empirical support for displacement of the prevailing approach; after all, what they were doing was apparently adequate yesterday, so why was it now unacceptable?

5 Implications for Practice

Managers have a clear interest in improving situated performance and are therefore interested in improvements to the state of organisational knowledge through a learning process. What this study has shown is that there are two distinct components of this organisational knowledge that must be addressed simultaneously if the expected result is to be achieved. This is made more difficult by the fact that the two components cannot be addressed in the same way. Improvements in performative capability are readily achieved by conventional means: training, resources, management support, metrics and incentives. These methods however, are ineffective in achieving the displacement necessary for improvements in the ostensive component of organisational knowledge. A full examination of the means by which these "justified beliefs" persist or are changed is beyond the scope of this study, but the evidence is sufficient to suggest that there is a strong cultural component to beliefs and that they change slowly—if at all.

This tends to be particularly true if the existing beliefs have been validated by years of apparent success as was the case in all of the companies in this study. Under these conditions it seemingly requires a crisis in order to displace the current beliefs and open the door to the desired change. It has even been suggested that crises can be manufactured for this purpose (Kim 1998). A less draconian approach is to assume that, at least during a transition period, that one simply cannot count on the rank and file to "get it" and to interpret the problem correctly or to apply their expertise in the right way. This creates the somewhat counter-intuitive recommendation that we should focus less on results achieved and more on the activities and behaviours involved. A good example of this was observed in the third company studied. This company was judged to be the furthest along of the three in terms of Lean philosophy. One of the policies that had been put in place was that they would build exactly to order, no more and no less. On some days that meant setting up to produce one unit of a product variant. In spite of their Lean training, many of the employees were strongly opposed to this policy; they did not see how it could be worth the trouble involved and wanted very much to be able to run larger batches. It may or may not have been possible to change those beliefs over time, but in the short term the problem was avoided by telling people how to do their jobs and not allowing them to proceed in what they might have thought was the best manner.

This need to focus on behaviour rather than results is rooted in the fact that metrics tend to mask the underlying assumptions and can be manipulated to show results that are at odds with reality (Melnyk et al. 2010). This is essentially what happened in our setup time reduction example: what appeared to be excellent results masked the fact that the problem was not really being approached in the right way. As in that case, it is often possible to achieve a given result in a variety of ways or with different trade-offs. Naturally it is hoped and expected that appropriate choices will be made, but these are rarely spelled out. Instead, the use of metrics and incentives in performance management is heavily dependent on what (Hanson et al. 2011) called *informal alignment*—the extent to which employees have absorbed the prevailing perception of the problem and the appropriate range of options for addressing it. Attempting to change direction by changing the metrics is apt to result in behaviour that is at cross-purposes with the intended change.

To summarise, a manager attempting to foster organisational learning must develop an awareness of the two distinct components of organisational knowledge that must be addressed. This would be followed by an assessment of the gap that exists between the current and the desired states—on both dimensions. Where a gap exists in the ostensive aspect, a choice must be made whether to try to narrow the gap by changing minds—being mindful of the displacement issue and the cultural basis of justified belief—or to bridge the gap by dictating behaviour as was done in our example. Our evidence suggests that the latter is likely to be more effective, and this finding is echoed by Ettlie and Rosenthal (2008).

6 Implications for Research

The problems facing the researcher are in many ways similar to those faced by management in the sense that both are, to a degree, outsiders when it comes to understanding what is really going on in an organisation and must rely on standardised measures to tell the story. The procedural or performative aspects of organisational knowledge are much easier to observe and catalogue than the ostensive aspects and as a result researchers are often in the position of trying to explain situated performance as a dependent variable on the basis of the performative elements as independent variables. In light of the above discussion and the cited examples from the cases, it is clear that this creates a major missing variable problem. This has been very apparent in the work on Lean Production, where the procedural elements of Lean have proved to be of limited value in predicting the ultimate success of the initiative (Oliver et al. 1996; Lewis 2000).

In essence, this is a problem of interpreting variance in observed behaviour (Pentland 2003). Armed with the two-part view of organisational knowledge, we can see that variance can be good or bad: it can be the result of intelligent fine-tuning of the performance to suit the circumstances, or it can be an inability to follow best practices. As a result, we cannot know whether certain observable characteristics should be predictive of success or failure until we know something about the motivation behind them—the ostensive component of organisational knowledge. Absent that knowledge, a data set that contains examples of both types of variance can be expected to produce equivocal or non-significant results.

Unfortunately, this component is not readily accessible to conventional research techniques, particularly since the participants themselves may not be able to fully articulate the logic behind their actions (Cohen and Bacdayan 1994). Furthermore, there is no reason to assume that the logic that exists at the researcher's point of contact (typically the person filling out a survey or being interviewed) is the same as that where the work is being done (Rothaermel and Hess 2007). This state of affairs does not lead to easy solutions, but does suggest some recommendations. The first, obviously, is awareness. When we recognise that situated performance is dependent on two very distinct components of organisational knowledge, we are better equipped to explain what we see. Secondly, we must design our research methods to ask the right questions. While we may not get total clarity, we must start to see organisations as collections of individuals acting within a cultural context whose characteristics are at least notionally discoverable. Finally, these studies have made it clear that if we want to better understand how organisations learn and function, we are going to have to get more deeply embedded in them. Ethnography is our guide here and teaches us that cultures are not readily understood from the outside.

7 Conclusion

The value of knowledge as an asset of an organisation is such that a great deal of attention has been paid to its "management". Unfortunately, the language of knowledge management has created an image of organisational knowledge as having an independent existence such that it can be stored, retrieved, transmitted, absorbed and replicated. A pragmatic view of knowledge (situated performance) suggests that none of these terms are strictly applicable and that knowledge consists of two parts that are incommensurate, but must both exist in the same place at the same time.

As a result, we cannot manage knowledge directly; rather we must manage its component parts in order to create new and unique instances of organisational knowledge as required. Managing the performative aspects is quite well understood, and many examples of knowledge management in practice do just that through training and documentation. Managing the ostensive component is not only more difficult, as these cases have shown, but ultimately more important. Because the ostensive component provides the core of "justified belief", the mechanics of justification must be understood. Although this study has barely scratched the surface in this respect, it was clear that justified belief is rooted in, and specific to the cultural setting—to the extent that an individual might legitimately hold different beliefs in different settings. This promises to be the next frontier in knowledge management—one that is not based on artificial intelligence and information technology, but instead in the behavioural sciences.

References

A. Amin, P. Cohendet, Organisational learning and governance through embedded practices. J. Manag. Gov. **4**(1–2), 93 (2000)

S.D. Amundson, Relationships between theory-driven empirical research in operations management and other disciplines. J. Oper. Manag. **16**(4), 341–359 (1998)

C.A. Anderson, J.J. Lindsay et al., Research in the psychological laboratory: truth or triviality? Curr. Dir. Psychol. Sci. **8**(1), 3–9 (1999)

C. Argyris, D.A. Schön, *Organizational Learning: A Theory of Action Perspective* (Addison-Wesley Publishing Company, Reading, 1978)

C.I. Barnard, *The Functions of the Executive* (Harvard University Press, Cambridge, 1938)

J. Barney, Firm resources and sustained competitive advantage. J. Manag. **17**(1), 99 (1991)

C.A. Bartel, R. Garud, in *Narrative Knowledge in Action: Adaptive Abduction as a Mechanism for Knowledge Creation and Exchange in Organizations*, ed by M. Easterby-Smith, M.A. Lyles. The Blackwell Handbook of Organizational Learning and Knowledge Management (Blackwell Publishing, Oxford, 2003)

K.A. Bates, E.J. Flynn, Innovation history and competitive advantage: a resource-based view analysis of manufacturing technology innovations. Acad. Manage. Proc. **1995**(1), 235–239 (1995). doi:10.5465/ambpp.1995.17536502

R.A. Bettis, M.A. Hitt, The new competitive landscape. Strateg. Manag. J. **16**, 7 (1995)

M.D. Cohen, Individual learning and organizational routine: emerging connections. Organ. Sci. **2**(1), 135–139 (1991)

M.D. Cohen, P. Bacdayan, Organizational routines are stored as procedural memory: evidence from a laboratory study. Organ. Sci. **5**(4), 554–568 (1994)

W.M. Cohen, D.A. Levinthal, Absorptive capacity: a new perspective on learning and innovation. Adm. Sci. Q. **35**, 128–152 (1990)

D.J. Collis, C.A. Montgomery, *Corporate Strategy: A Resource-Based Approach* (Irwin/McGraw-Hill, Boston, 1998)

R.M. Cyert, P. Kumar et al., Information, market imperfections and strategy. Strateg. Manag. J. (1986–1998) **14**(SPECIAL ISSUE), 47 (1993)

R.M. Cyert, J.G. March, *A Behavioral Theory of the Firm* (Prentice-Hall, Englewood Cliffs, 1963)

T.H. Davenport, L. Prusak, *Working Knowledge: How Organizations Manage What They Know* (Harvard Business School Press, Boston, 1998)

P. Dennis, *Lean Production Simplified: A Plain Language Guide to the World's Most Powerful Production System* (Productivity Press, New York, 2002)

J. Dewey, *Democracy and Education: An Introduction to the Philosophy of Education* (Collier-MacMillan, London, 1916)

J. Dewey, A.F. Bentley, *Knowing and the Known* (Beacon Press, Boston, 1949)

P.F. Drucker, *Post Capitalist Society* (Harper Business, New York, 1993)

A.C. Edmondson, A.B. Winslow et al., Learning how and learning what: effects of tacit and codified knowledge on performance improvement following technology adoption. Decis. Sci. **34**(2), 197 (2003)

L.M. Ellram, The use of the case study method in logistics research. J. Bus. Logist. **17**(2), 93–138 (1996)

M.L. Emiliani, *Real Lean: Understanding the Lean Management System* (The Center for Lean Business Management, LLC, Wethersfield, 2007)

J.E. Ettlie, S.R. Rosenthal, Service innovation in manufacturing (Final Report: National Science Foundation Grant No.0453694) (2008)

M.S. Feldman, B.T. Pentland, Reconceptualizing organizational routines as a source of flexibility and change. Adm. Sci. Q. **48**(1), 94 (2003)

M.S. Feldman, B.T. Pentland, Issues in empirical field studies of organizational routines. Working Paper (2005)

B.B. Flynn, S. Sakakibara et al., Relationship between JIT and TQM: practices and performance. Acad. Manag. J. **38**(5), 1325–1360 (1995)

R.M. Grant, Toward a knowledge-based theory of the firm. Strateg. Manag. J. (1986–1998) **17**(Winter Special Issue), 109 (1996)

R.W. Hall, *Zero Inventories* (Dow Jones-Irwin, Homewood, 1983)

J.D. Hanson, S.A. Melnyk et al., Defining and measuring alignment in performance management. Int. J. Oper. Prod. Manag. **31**(10), 1115–1139 (2011)

B. Hedberg, *How Organizations Learn and Unlearn*, ed. by P.C. Nystrom, W.H. Starbuck. Handbook of Organizational Design (Oxford University Press, Oxford, 1981), pp. 3–27

R. Henderson, I. Cockburn, Measuring competence? Exploring firm effects in pharmaceutical research. Strateg. Manag. J. (1986–1998) **15**(SPECIAL ISSUE), 63 (1994)

B.A. Hoey, What is ethnography? (2011), http://www.brianhoey.com/General%20Site/general_defn-ethnography.htm. Accessed 15 June 2011

S.D. Hunt, *Modern Marketing Theory: Critical Issues in the Philosophy of Marketing Science* (South-Western Publishing Co, Cincinnati, 1991)

A. Kaplan, *The Conduct of Inquiry, Methodology for Behavioral Science* (Chandler Publishing, San Francisco, 1964)

L. Kim, Crisis construction and organizational learning: capability building in catching-up at Hyundai motor. Organ. Sci. **9**(4), 506–521 (1998)

B. Kogut, U. Zander, What firms do? Coordination, identity, and learning. Organ. Sci. **7**(5), 502 (1996)

D. Leonard-Barton, Core capabilities and core rigidities: a paradox in managing new product development. Strateg. Manag. J. **13**(SPECIAL ISSUE), 111 (1992)

B. Levitt, J.G. March, Organizational learning. Ann. Rev. Sociol. **14**, 319–340 (1988)

M.A. Lewis, Lean production and sustainable competitive advantage. Int. J. Oper. Prod. Manag. **20**(8), 959–978 (2000)

J.G. March, H.A. Simon, *Organizations* (Wiley, New York, 1958)

J. Martin, *Organizational Culture: Mapping the Terrain* (Sage, Thousand Oaks, 2002)

S.A. Melnyk, R.J. Calantone et al., Short-term action in pursuit of long-term improvements: introducing Kaizen events. Prod. Inventory Manag. J. **39**(4), 69–76 (1998)

S.A. Melnyk, J.D. Hanson et al., Hitting the target but missing the point: resolving the paradox of strategic transition. Long Range Plan. **43**(4), (2010)

M.B. Miles, A.M. Huberman, *Qualitative Data Analysis: An Expanded Sourcebook* (Sage, Thousand Oaks, 1994)

D. Miller, Configurations revisited. Strateg. Manag. J. (1986–1998) **17**(7), 505 (1996)

I.I. Mitroff, R.O. Mason, On evaluating the scientific contribution of the Apollo moon missions via information theory: a study of the scientist–scientist relationship. Manag. Sci. **20**(12), 1501–1513 (1974)

A.S. Mukherjee, M.A. Lapre et al., Knowledge driven quality improvement. Manag. Sci. **44**(11), S35 (1998)

J. Nahapiet, S. Ghoshal, Social capital, intellectual capital, and the organizational advantage. Acad. Manag. Rev. **23**(2), 242 (1998)

R.R. Nelson, Why do firms differ, and how does it matter? Strateg. Manag. J. (1986–1998) **12**(SPECIAL ISSUE), 61 (1991)

I. Nonaka, A dynamic theory of organizational knowledge creation. Organ. Sci. **5**(1), 14 (1994)

I. Nonaka, H. Takeuchi, *The Knowledge-Creating Company* (Oxford University Press, New York, 1995)

T. Ohno, *Toyota Production System: Beyond Large-Scale Production* (Productivity Press, New York, 1988)

N. Oliver, R. Delbridge et al., Lean production practices: international comparisons in the auto components industry. Br. J. Manag. **7**(1), 29–44 (1996)

W.J. Orlikowski, Knowing in practice: enacting a collective capability in distributed organizing. Organ. Sci. **13**(3), 249–273 (2002)

E.T. Penrose, *The Theory of the Growth of the Firm* (Oxford University Press, Oxford, 1959)

B.T. Pentland, Organizing moves in software support hot lines. Adm. Sci. Q. **37**(1992), 527–548 (1992)

B.T. Pentland, Grammatical models of organizational processes. Organ. Sci. **6**(5), 541 (1995)

B.T. Pentland, Sequential variety in work processes. Organ. Sci. **14**(5), 528 (2003)

B.T. Pentland, M.S. Feldman, Organizational routines as a unit of analysis. Ind. Corp. Change **14**(5), 793–815 (2005)

M. Polanyi, *Personal Knowledge: Towards a Post-Critical Philosophy* (University of Chicago Press, Chicago, 1958)

C.K. Prahalad, G. Hamel, The core competence of the corporation. Harv Bus Rev **68**(3), 79 (1990)

E. Rosch, B.B. Lloyd (eds.), *Cognition and Categorization* (Lawrence Erlbaum Associates, Hillsdale, 1978)

F.T. Rothaermel, A.M. Hess, Building dynamic capabilities: innovation driven by individual, firm, and network-level effects. Organ. Sci. **18**(6), 898–921 (2007)

R. Sabherwal, I. Becerra-Fernandez, An empirical study of the effect of knowledge management processes at individual, group, and organizational levels*. Decis. Sci. **34**(2), 225 (2003)

E.H. Schein, How can organizations learn faster? The challenge of entering the green room. Sloan Manag. Rev. **34**, 85–92 (1993)

J.-C. Spender, Making knowledge the basis of a dynamic theory of the firm. Strateg. Manag. J. (1986–1998) **17**(Winter Special Issue), 45 (1996)

W.H. Starbuck, Learning by knowledge-intensive firms. J. Manag. Stud. **29**(6), 713 (1992)

D.J. Teece, G. Pisano et al., Dynamic capabilities and strategic management. Strateg. Manag. J. (1986–1998) **18**(7), 509 (1997)

J.P. Walsh, Managerial and organizational cognition: notes from a trip down memory lane. Organ. Sci. **6**(3), 280 (1995)

K.E. Weick, *The Social Psychology of Organizing* (McGraw-Hill, New York, 1979)

B. Wernerfelt, A resource-based view of the firm. Strateg. Manag. J. **5**(2), 171 (1984)

O.E. Williamson, The economics of organization: the transaction cost approach. Am. J. Sociol. **87**(3), 548–577 (1981)

R.K. Yin, *Case Study Research Design and Methods* (Sage Publications, Thousand Oaks, 1994)

Chapter 10
Behavioral Decision-Making and Network Dynamics: A Political Perspective

Francesco Zirpoli, Luisa Errichiello and Josh Whitford

Abstract The blurring of organizational boundaries and the adoption of networks as a prominent form of *governance* have largely contributed to reinforcing inter-dependence between internal and external organizational networks as well as between formal and informal ties. This chapter tries to broaden existing theoretical models in order to explain the behavioral decision-making process of the firm and how it is shaped by the complex and interactive dynamics of these networks. The theoretical perspective employed in the chapter suggests that a firm's behavior is influenced by organizational politics. Although this actually does not constitute a fresh perspective within organizational and management studies, in this chapter it is revamped and widened in light of the mentioned changes within and across firms' organizational boundaries. The starting point of our discussion is March's seminal work (March in Journal of Politics 24(4):662–678, 1962) and his model of "the business firm as a political coalition". Subsequently, drawing also on later *organizational politics* literature we show the limits and opportunities of adopting such an imagery not only for the traditional business firm but also for the con-temporary network organization: through it we can improve our understanding of how organizational boundaries are defined today, why company leaders choose the

F. Zirpoli (✉)
Department of Management, University Ca' Foscari of Venice, San Giobbe,
Cannaregio 873 30121 Venezia, Italy
e-mail: fzirpoli@unive.it

L. Errichiello
Institute for Service Industry Research (IRAT), The National Research Council
of Italy (CNR), Via Michelangelo Schipa, 91 80122 Naples, Italy
e-mail: l.errichiello@irat.cnr.it

J. Whitford
Department of Sociology, Columbia University, 606 West 122nd Street,
MC 9649, New York, NY 10027, USA
e-mail: jw2212@columbia.edu

I. Giannoccaro (ed.), *Behavioral Issues in Operations Management*,
DOI: 10.1007/978-1-4471-4878-4_10, © Springer-Verlag London 2013

strategies they choose, and how and why those strategies are (or are not) implemented. In order to explain patterns of organizational behavior in a world of blurred-but existent firm boundaries we finally draw on a more recent sociological literature on social movements that also highlights for "patterns of mobilization distinct from both lines of formal authority and the personal ties of informal organization" (Clemens, Where Do We Stand? Common Mechanisms in Organizations and Social Movements Research, in Davis G, McAdam D, Scott WR, Zald M (eds) Social movements and organization theory. Cambridge University Press, Cambridge, p. 356, 2005). Indeed, such a literature recognized the central role of networks, their evolutionary dynamics, and interactions between the internal and the external and between the formal and the informal.

1 Firm Boundaries, Networks, and Politics

Over the past few decades firm boundaries have rapidly and significantly changed, actually becoming more and more blurred. Indeed, identifying their exact contours has increasingly revealed itself as a challenging task for firms in a number of industries, such as automotive (Dyer 1996; Helper et al. 2000), pharmaceutical (Powell 1998), biotechnology (Powell 1996, 1998), electronics, (Sturgeon 2002), mechanical (Herringel 2004; Whitford 2005), and software (Lerner and Tirole 2002).

Within the realm of economic globalization, we assisted toward a number of paradigmatic shifts: the emergence of new Japanese-based models of industrial organization, the growing importance attached to services and knowledge (Gadrey and Gallouj 2002), high efficient markets of technology (Arora et al. 2001), a marked trend toward specialized technological knowledge (Langlois 2003). There is no question that all these changes have deeply affected firm strategies and the overall architecture of industrial organization: vertical disintegration, outsourcing, and collaboration-based modes of organizing emerged as the dominant strategic options to compete and survive in turbulent environments and uncertain markets. In most cases, the choice made by firms to externalize production or peripheral activities has its roots in their will to maintain a competitive position through lower prices and shorter *lead-time*, greater strategic and operational flexibility, higher quality of products and services, as well as the access to new sources of knowledge and specialized competencies. As a consequence, firms deliberately started to build vertical networks of production and innovation. At the same time, new and complex relationships with external actors, especially suppliers, gradually developed from those initially established as purely formal structures.

Outsourcing has not limited its effect to a change to the internal organization of the firm. In fact, by choosing to outsource some of their activities, firms have also contributed—often unconsciously—to modify the organizational structure of external organizations, which have gradually achieved a prominent role in production and innovation processes. Consequently, what had been conceived by the management as a simple outsourcing strategy has turned, over the years, into a mutual-dependent collaboration or an ambiguous bilateral relationship (Helper et al. 2000) where organizational boundaries have actually become *blurred-but-existent.*

Jointly with dramatic changes in organizational boundaries and the rapid diffusion of new collaborative arrangements that are dissimilar both to markets and hierarchies, the organizational and managerial literature has gradually recognized the network as a "distinctive form of coordinating the economic activity" (Powell 1990, p. 301). In this regard, an extensive research tradition based in sociological, organizational, and strategic domains has notably discussed the several advantages deriving from this form of *governance*, and tried—simultaneously—to shed light on its specific characteristics and underlying mechanisms.

With regard to the first theme, many studies have hitherto shown the multiple advantages inherent in various forms of *networking*. In this respect, it is worth highlighting that, by relying on networked arrangements, firms could draw on financial, market, and knowledge resources, conduct joint *problem-solving* activities, and sustain learning and innovation processes (Gulati and Gargiulo 1999; Lorenzoni and Lipparini 1999; Dyer and Nobeoka 2000; Powell et al. 1996). In addition, many authors (Carruthers and Uzzi 2000; DiMaggio 2001; Podolny and Page 1998) have highlighted how in facing uncertain markets and technologies that network forms of governance tend to appeal to actors when "both markets and environments change frequently and there is a premium on adaptability" (Smith-Doerr and Powell 2005, p. 380).

As for the functioning of network organizational forms, remarkable efforts have been done in this direction to identify the contours of a distinctive "embedded" logic of exchange (Helper et al. 2000; Jones et al. 1997; Podolny and Page 1998), at the intersection between transactional and relational modes. Indeed, although the choice to adopt the network as a form of governance is mainly based on the evaluation of related business opportunities, its emergence and consolidation tend to be highly influenced by pre-existent social relationships and by their evolution over time (Azoulay 2003; Helper et al. 2000; Podolny and Page 1998).

Although the notion of *network governance* acknowledges—next to the legal— the social nature of firms' relationships (Jones et al. 1997), studies on inter-organizational networks (i.e., subforniture, strategic alliances, research consortia, joint ventures) tend to mainly focus on firms or well-defined organizational units as well as on formal ties (Hagedoorn and Duysterns 2002; Smith-Doerr and Powell 2005). On the contrary, within intra-organizational studies, the *network form* is mostly viewed as a kind of organizational structure next to the multidimensional and functional one (Bartlett and Ghoshal 1993; Hedlund 1994; Miles and Snow 1995; Nohria and Eccles 1992), whereas the focus is primarily on informal ties (Cross et al. 2002; Hansen et al. 2001; Nahapiet and Ghoshal 1998; Podonly and Baron 1997). Finally, it is worth highlighting that there are only few studies on *network governance* where the analytical focus is on the interdependencies between formal and informal networks or alternatively between external and internal linkages. Falling within this tradition we find some studies that seriously investigate how the internal organization of potential collaborators fundamentally affects their ability to build alliances with external actors (Azoulay 2003; Helper, et al. 2000; Kristensen and Zeitlin 2005; Lorenzoni and Lipparini 1999). And, at least to our knowledge, no attention has been paid to effects that run in the other direction, that is, to the implications that patterns of external relations have for the internal organization of potential collaborators and the interactive dynamics developing at the boundaries of the firm (for an exception see Parmigiani and Mitchell 2009). Nevertheless, the importance of taking into direct consideration the above mechanisms appears evident in light of recent economic and organizational changes that have affected firms and their boundaries. Indeed, although the two cited literature traditions have ostensibly found a peaceable division of labor, they have in fact deprived themselves of a dialog that is necessary if they are to passably theorize processes of organizational change and adaptation in a world transformed by radical outsourcing. This is obviously a problem given that adaptation processes are key concerns for both the evolutionary theory (Nelson and Winter 1982) and the behavioral theory of the firm (Cyert and March 1963), and therefore also for the massive literatures built upon those theories (see Gavetti and Levinthal 2004).

Furthermore, the literature on network organizational forms shows a serious methodological drawback (Smith-Doerr and Powell 2005), that is to say the predominance of statistical analyses, mainly focusing on the structural elements of relationships and often relying on sophisticated analysis techniques and graphic visualization tools. As a matter of fact, such an analysis does not allow to understand important aspects related to organizational networks, such as the antecedents for their creation, their evolution over time, as well as the consequences of such evolutionary dynamics on firms' strategies and adaptation capabilities.

The blurring of organizational boundaries and the adoption of networks as a prominent form of *governance* have largely contributed to reinforcing interdependence between internal and external organizational networks as well as between formal and informal ties. The outlined gaps highlight the chance to broaden existing theoretical models in order to explain the behavioral decision-making process of the firm (that is its strategic decisions and the consequences deriving from these decisions for the organization) and how it is shaped by the complex and interactive dynamics of these networks.

The theoretical perspective suggested here that firm's behavior is influenced by organizational politics. Although this actually does not constitute a fresh perspective within organizational and management studies, in this chapter it is revamped and widened in light of the mentioned changes within and across firms' organizational boundaries. The starting point of our discussion is March's seminal work (1962) and his model of "the business firm as a political coalition". Subsequently, drawing also on later *organizational politics* literature we show the limits and opportunities of adopting such an imagery not only for the traditional business firm but also for the contemporary network organization: through it we can improve our understanding of how organizational boundaries are defined today, why company leaders choose the strategies they choose, and how and why those strategies are (or are not) implemented. In order to explain patterns of organizational behavior in a world of blurred but existent firm boundaries we finally draw on a more recent sociological literature on social movements that also highlights for "patterns of mobilization distinct from both lines of formal authority and the personal ties of informal organization" (Clemens 2005, p. 356). Indeed, such a literature recognized the central role of networks, their evolutionary dynamics, and interactions between the internal and the external and between the formal and the informal.

2 Conflict, Coalitions, and Power Within the Organization and the Network

In line with previous studies in organizational theory (March and Simon 1958; Thompson 1961), James March's seminal (1962) analysis, *The business firm as a political coalition*, depicts the business organizations understood by the Carnegie school as a conflicting socio-political system. The underlying attributes of a *conflict system* are: (1) the existence of basic units with consistent preference orderings; (2) their mutual inconsistency relative to the resources of the system that is "the most preferred states of all elementary units cannot be simultaneously realized" (March 1962, p. 663). The author speculates on their theoretical assumptions arguing that "the preference ordering of the subsystem (which constitutes the elementary unit) is casually antecedent, and independent of, the decisions of the larger system" or, alternatively, that "variation in system behavior due to conflict within the subsystem is trivial because of scale differences between the conflict within the subsystem on the one hand and conflict among subsystems on the other" (March 1962, p. 664).

Extant classical theories presume to resolve conflict by "simple payments and agreement on a superordinate goal" (March 1962, p. 674). Long-run profit maximization or leveraging incentives (i.e., employment contracts or payments) are a case in point. However, March considers this theory wrong, as it overlooks the complexity of an organization as well as the plurality of individuals and interests involved. Accordingly, he models the business firm as a conflict system where decisions on the allocation of resources are made by coalitions of interest groups having a certain potential control over the system. In such a system, demands will be made on executives by participants to coalitions whose cooperation or concession affects the firm's competitive position. In fact, March claims that the executive of the firm can be seen as a "political broker", that "the composition of the firm is not given but negotiated and its goals are not given but bargained" (March 1962, p. 672) so that a number of coalitions will be possible at any given point in time.

All these concepts are taken for granted today in studies on organizational politics which by drawing on and extending March's innovative claims, have gradually developed a rich corpus of theories on organizational and inter-organizational decision-making processes. And, in fact, concepts such as politics, influence and power are highly recurrent in this well-established research tradition (Bower and Doz 1979; Elg and Johansson 1997; Pettigrew 1977; Pfeffer and Salancik 1978; Pfeffer 1981).

In Pettigrew (1977), for example, a political perspective is introduced to conceptualize the strategy of the firm as a process of conflict resolution between the contrasting requests exposed by different individuals or groups. Beyond studies on single organizations (Bower 1970; Burgelman 1983; Pettigrew 1977; Pfeffer 1981), in other contributions (Elg and Johansson 1997; Leblebici and Salancik 1982; Pfeffer and Salancik 1978; Salancik 1986) it is widely acknowledged that business networks, similarly to single firms, are systems of conflicting interests in which power structure and political action within and between member firms remarkably influence their respective decision-making processes.

Although the existence of political coalitions and of interest-based behavior are key assumptions in Cyert and March's work (1963) theoretical speculation actually remained limited to conflict resolution mechanisms, while scarce attention is attached to the equally important processes that lead to their formation (Pettigrew 1973).

Primary aiming to fill this gap, the literature on strategy-making processes (Bower 1970; Burgelman 1994; Pettigrew 1973) has elaborated political models where individuals/groups' interests and incentive mechanisms essentially guide the mobilization of "power resources" and the control of informational flows, ultimately influencing the internal dynamics of the strategy-making process. Concepts such as resources, power, and structure are all central to those research streams that adopt a political lens to explore and explain organizational and inter-organizational decision-making processes (Krackhardt 1990; Pfeffer and Salancick 1978; Pfeffer 1981). In these contributions the structural position of the actors—individuals, groups, or firms—largely influences their capacity for exerting and affecting the mobilization of resources within and outside the organization in order to control its decision-making processes.

3 The Business Firm as a Political Coalition: Limits and Opportunities of a Consolidated Framework

The dramatic changes that happened in organizational models, increasingly based on outsourcing and networked innovation, along with more permeable and fluid organizational boundaries, set the stage to review the seminal work on March. Indeed, we retain many of the key concepts strongly anchored to the classical view of the firm as a political coalition. However, by explicitly addressing the firm's evolution toward new governance modes, we also show the limits that—in this respect—such contribution shares with most traditional studies on organizational politics. Our final goal is to revamp March's seminal contribution through framing the image of what we called "the network as a political coalition". Definitively, we want to keep faithful to a "behavioural" tradition (Cyert and March 1963) through looking at what "actually" happens in organizations rather than what is ideally expected to happen (Pettigrew 1973). To this end, we need to explicitly consider the *blurred-but-existent* nature of firm's boundaries when we try to understand the role played by politics in shaping a firm's behavior and its network dynamics.

We take our theoretical starting point from March's analysis (1962) since many of the ideas presented in that article and partially resumed in Cyert and March (1963) have come essentially to define what it means to understand an organization in political terms. Specifically, we retain March's expectation that demands will be made on executives by participants to coalitions whose cooperation or concession affects the firm's competitive position, that a number of coalitions will be possible at any given point in time and, as March, we reject the assumption that conflict problems can easily be solved by "simple payments and agreement on a super-ordinate goal". Accordingly, we assume that, at any given point in time, a number of political coalitions could exist, each with its own interests, demands, and influence on the political broker (i.e., the executive of the firm) and that cooperation or concession by participants to coalition finally affect the firm's competitive position. As in his seminal work, we recognize that the sorts of "elementary units" likely to be of interest which would include not only individuals, but also work groups, departments, functional areas, and other such things that are themselves conflict systems.

At this stage our analysis departs from March's analysis, since we consider poorly realistic the assumption that, in analyzing the demands that the elementary units, i.e., the subsystems, place on the system are either independent of the decisions of the larger system, or that "variation in system behavior due to conflict within the subsystem is trivial because of scale differences between the conflict within the subsystem on the one hand and conflict among subsystems on the other" (March 1962, p. 664). Indeed, according to this view, such elementary units can always be univocally determined so that they do not significantly affect the system's behavior.

This proposition largely remained unchanged in subsequent work, largely influenced by March's argument. As a matter of fact, later organizational studies recognized the central role played by politics, in terms of power and influence, in shaping interorganizational dynamics as well as the importance of existing inter-firm organizational coalitions. In their analysis, however, authors retain the simplifying assumption made by March (1962) in order to show that organizations depend on the environment for their survival and that those dependencies typically take the form of relationships between organizations understood as essentially bounded but interacting units with a clear set of preferences and interests (Elg and Johansson 1997; Leblebici and Salancik 1982; Salancik 1986).

However, in light of the well-documented changes in the organization of production, relying on the assumption that outcomes of interest are relatively unaffected by a blurring of boundaries between desired units of analysis seems quite unrealistic since it badly reflects the new organizational and competitive scenario. Although not obviating the relevance of a coalitional imaginary in the analysis of organizational behavior, we have to articulate such a frame through considering the quantitative intensity, complexity, and frequency of individual-to-individual ties at multiple levels, within and across the formal boundaries of the organization. In fact, the blurring-but-existence of organizational boundaries means that it is increasingly difficult to identify unambiguously functions, roles, routines as well as their attribution to well-established formal units, while incentive mechanisms are not always directly controlled by formal roles within the organization. Not only cannot units such as groups, functional units, communities of practice, etc. be assumed as "elementary" in nature, but also the existence of *one-to-one* relations and a shared set of preferences within each unit have to be seriously questioned.

In our argument, coalitions are understood as Cyert and March (1963, p. 39) and the ensuing literature did, i.e., as temporary alliance among some subset of the involved parties. Simultaneously, we expect that those coalitions constrain executives and their choices. However, we need space for the possibility that those coalitions will cut across organizational boundaries in an interactive way. Specifically, we have to consider how the recourse to the network as a form of governance and the blurring-but-existent nature of organizational boundaries shape conflict dynamics within the subunits themselves, the formation of cross-firm political coalitions as well as the increasing interdependence between the demands expressed by individuals and subgroups (not always belonging to the firm) and the strategic decisions taken at executive level. In this respect, the ability of particular actors (or groups of actors) in one organization to achieve their goals will often depend considerably not just on actors in their own organizations but also on actors (or groups of actors) in other organizations. And we should therefore expect that demands placed by suppliers on the executive will be far less and less independent of decisions made by that executive than they were when much of the contemporary groundwork in the literature on organizational politics was laid. In other words, in order to understand how power dynamics helps to explain organizational behavior we need to acknowledge and analyze the *many-to-many* relationships among units and the social interactions among individuals or subsets within units.

With these remarks, we can move toward a model of firm behavior where the political dimension serves the function to explain what happen within and across organizational boundaries. In order to articulate the new imaginary of the "network as a political coalition", it is needed to explain why actors (or groups of actors) with conflicting interests ultimately enrol in a given coalition rather that in another, and how their enrolment in that coalition impacts the demands they place on the center as well as that the center (and periphery) place on them. For this purpose, the role of *cross-firm* formal and informal networks has to explicitly be addressed. An important step in this direction is the abandonment of a static view of the network, focused on resources, structures, and power (Leblebici and Salancik 1982; Salancic 1986). In this respect, we draw on a more recent research stream in this tradition (Elg and Johansson 1997; Hardy 1996; Hardy et al. 2003; McLoughlin et al. 2001) that, on the contrary, criticizes an excessive emphasis on the "dependence on resources" as well as on the role that power distribution plays in ensuing control over resources. According to these studies, an exclusive analytical and static focus on resource dependence is likely to put in the background the role played by dynamic interactions among actors in affecting the formation of coalitions (internally and externally the firm) as well as the capacity of some actors to exploit resources and structures in pursuing their own goals. In attempting to put on the foreground these dynamics and explicitly including the evolutionary dimension in a political network analysis, we will show (in the next paragraphs) how the most recent literature on social movements can fruitfully contribute to enriching traditional organizational studies, not only at the micro and macro levels (as already have been discussed in the relevant literature), but also and especially at the meso-level, where it provides a conceptual apparatus that flows easily across shifting organizational boundaries and as such can be applied to analyze an organizational network from a political perspective.

4 The Contribution of Social Movements Literature

Organizational studies and social movements literature share a number of core concepts and modes of analysis (Campbell 2005; Davis and Zald 2005; McAdam and Scott 2005). However, for many decades, they have been treated as distinct literary traditions and developed according to substantially independent paths. Over the past few years the fruitful opportunities of cross-fertilization between such disciplinary areas have been systematically considered and, in particular, organizational scholars have started to discover the interesting theoretical cues and application potential that social movement literature has to offer to their studies (Davis et al. 2005; Davis and McAdam 2000; Davis and Thompson 1994; Rao et al. 2000). More specifically, looking at mechanisms underlying the development and change of social movements, we are able to understand organizational change as well as adaptation processes to environmental changes (Davis et al. 2005). Indeed, mobilization mechanisms, that are the analytical focus of social movements studies, remain substantially unchanged also in different contexts and times (Campbell 2005). This does mean that they can potentially be applied also to understand organizational phenomena.

As highlighted by McAdam and Scott (2005), the two somewhat divergent approaches mostly adopted in the study on social movements and that we consider particularly relevant in this work are those relying on the key concepts of resource mobilization (Edwards and McCarthy 2004; Gamson 1975; Zald and Berger 1978) and political processes (Tilly 1978). Scholars crafting resource mobilization stress the central role of power and politics, both within the organization and in its relation to the environment (McAdam and Scott 2005), so emphasizing the key role of organizational structures and processes in recruiting people, acquiring resources, and disseminating information (Campbell 2005). Accordingly, these elements, identifiable with what McAdam et al. (1996) defined "mobilizing structures", actually represent a key building-block of social movements. When embracing a political process perspective, the analytical focus shifts on "political opportunities" and constraints on social movements and then on those environmental factors that hold down, facilitate, and structure collective action (McAdam et al. 1996).

Next to mobilizing structures and political opportunities, McAdam et al. (1996) identified a third broad factor shaping the emergence and development of social movements: *framing* processes, i.e. "collective processes of interpretation, attribution and social construction that mediate between opportunity and action" (McAdam et al. 1996, p. 2) since they allow to interpret political opportunities and, accordingly, to decide which is the best way to achieve own goals.

The concept of *frame*, originally adopted by Goffman (1974), has been extensively employed in the research on political sociology, particularly for the study on social movements and collective action (Benford and Snow 2000). In this context, the term was used to indicate "metaphors, symbols and cognitive cues that cast issues in a particular light and suggest possible ways to respond to these issues" whereas "framing involves the strategic creation and manipulation of shared understandings and interpretations of the world, its problems, and viable courses of action" (Campbell 2005, p. 49). Therefore, a cognitive *frame* is the lens through which the actors perceive, interpret, and understand reality. The frame "acts as a boundary that keeps some elements in view and other out of view" thus conveying "what is or is not important by grouping certain symbolic elements together and keeping others out" (Williams and Benford 1996, p. 3). Ass a "meaning constructor", *framing* can be then viewed as the process through which the interpretive lens of reality is created; it is dynamic in nature, involves meaning construction and interaction mechanisms and focusing on the role of *agency* (Benford and Snow 2000). Framing refers to a signifying work, that is to the "processes associated with assigning meaning to or interpreting relevant events and conditions in ways intended to mobilize potential adherents and constituents, to garner bystander support, and to demobilize antagonists" (Benford 1997, p. 416). According to this perspective, when new political opportunities show up, actors taking part in a social movement—supporters or opponents of well-defined *frames*—carry out *framing* activities to "mobilize" other people toward a given point of view or interpretive lens of reality (i.e., a collective *frame*), thus leading, through their interaction, to the formation of political coalitions (Snow et al. 1986). The so-called *framing practices* can be in the form of discourse, consisting of dialogs, conversations and written communications, or strategic, when the goal is represented by the alignment/realignment of interests and collective frames to those of their supporters (Snow et al. 1986; Benford and Snow 2000). These processes are frequently contested and negotiated, not always under the tight control of an elite and not always yielding the desired results (Benford and Snow 2000).

In these practices the notion of "resonance" (Snow and Benford 1988) assumes a central role. Used to design the efficacy of a given frame during the mobilization process, it is primarily affected by two factors, i.e., the frame reliability and its relevance with regard to a given target (Benford and Snow 2000).

In helping to understand the network firm as a political coalition, the contribution of social movement theory is not confined to the concepts of mobilization processes, political opportunities, or framing practices. Indeed, it is important to highlight that in these studies, a key role in mobilization processes, both at the individual and inter-organizational level is played by the network (Diani 2003). Accordingly, through a careful examination of "how" the network is treated in social movements literature we can gain some important insights to understand the evolutionary dynamics underlying social networks, i.e., how they emerge and develop both formally and informally, both within and across the firm's boundaries, highly shaped by politics, power, and coalition formation. In studies on social movements, networks play a key role in mobilizing individuals toward collective action both in the early stages, where individual identities are built or consolidated and a potential for participation is created, and in the final phases, where preferences and perceptions are shaped and individuals are engaged in collective action (Passy 2003). In other words, at the individual level, network serves the functions of (Passy 2003):

- *Socialization*: through social interaction, networks convey meanings that build and solidify identities, shape actor's cognitive frames, enabling them to interpret social reality and define a set of actions, then preparing them for collective participation;
- *Structural connection*: networks play a mediating role by connecting prospective participants to an opportunity for mobilization and enabling them to convert their political consciousness into action;
- *Decision-making processes*: as, through social interaction, individual perceptions and interests and then the decision to join collective action are influenced by the action of other participants.

At the collective and organizational levels, networks serve as mobilizing structures that shape and constrain people's behavior and opportunities for action (Campbell 2005). Indeed, it is through social relationships that new models, concepts, and practices diffuse and become part of an organization or movement's repertoire and, therefore, become available for use in framing. Furthermore, they help to identify the sources of collective support for mobilization (Campbell 2005), facilitate the negotiation of shared goals as well as the production and diffusion of information, i.e., all those activities that are essential to any kind of coalition (Diani 2003). Networks are viewed as the channels for carrying out framing activities, thus favoring or impeding the circulation of well-defined meanings and cognitive frames.

Within social movements theory, the observation of *cross-firm* networks allows a clearer understanding of the criteria that guide the mobilization processes at individual, group, organizational, and inter-organizational levels and help to explain their choices to form or sustain occasional or permanent allies. In this regard, it is worth highlighting that it is just in the attempt to identify the factors at the basis of coalition and alliance formation at inter-organizational level that studies on social movements and organizational sociology have come to share a common goal (e.g., Gulati and Gargiulio 1999; Podonly and Page 1998).

The emerging organizational literature based on social movement theory was initially focused on a micro level of analysis, and only recently, it has begun to branch out increasingly into macro investigations (Clemens 2005; Davis and Zald 2005). In the first case, the unit of analysis is the single "focal" organization (Davis and McAdam 2000; Davis and Thompson 1994; Rao et al. 2000); in the latter, the study of mobilization processes is conducted at "field-level" (DiMaggio 1991; Fligstein and Maria-Drita 1996), that is to say on that aggregated system of actors, actions and relationships—different from both the single organization/ movement and a set of organizations/movements—so that among participants exist tangible reciprocities since they produce similar goods and services, i.e., they carry out interrelated activities (DiMaggio and Powell 1983; McAdam and Scott 2005).

As for studies conducted at the organizational level, the main contributions come from research on strategy making processes (Kaplan 2008; Levina and Orlikowski 2009). Here, the concept of *framing* is adopted to explain the relationship between cognition and the process of coalition formation. In an ethnographic study, through adopting a "practice lens" (Orlikowski 2000), focused on daily practices and routines, Kaplan (2008) examines, for instance, everyday organizational *strategy making* activities so as to identify the micro-mechanisms that interrelate cognitive *frames* and politics. For this purpose, the author widely draws on social movement theory, adopting concepts such as *framing practices*, *realignment processes, action mobilization, frame resonance* to elaborate a representation of strategy formulation processes where cognitive *frames* do not constitute static constraints (as in the predominant literature on social movements) but, on the contrary, are built up during daily practices through individual and group interaction, thus serving as a resource for collective action and the emergence of conflict. Emphasizing the centrality of "power" relationships within and across organizations, Levina and Orlikowski (2009), develop a model of power dynamics where the recourse to discursive resources (a kind of *framing practice*) coming from different institutional contexts allows to modify these relationships. In order to elaborate their model, the two authors analyze the everyday decisions taken in the field of a joint consultancy project, taking in account especially the conditions leading to discursive ambiguities, the modalities of resolution put in action by subjects through relying on discursive practices, and the consequences that the diversity among specific discursive practices have on power relationships within and across organizations.

To our knowledge, there have not been any efforts to date to understand processes of mobilization across an organizational network (i.e., at the meso level). However, the value of such a kind of contribution to the organizational theory is unquestionable, especially if we consider the renewed interest at this level of analysis in the study of social movements, where the relations between structure and agency need further investigation (Diani 2003). In this sense, our work does not simply draw on literature on social movements but it also aims at contributing to its theoretical development.

5 *Cross-Firm* Mechanisms and Mobilization Processes

Hitherto, we have recognized the fruitful insights that the study on social movements can provide to the study of organizations at meso-level. More specifically, we have showed how, through looking at the processes of inter- and intra-organizational mobilization in a world of blurred-but-existent organizational boundaries, we could gain a deeper understanding of network dynamics and in particular of the influence that power and politics have on coalition formation and firm's decision-making behavior. In this respect, we need now to analyze two key mechanisms of mobilization—*relational legacies* and *ideologies*—that have featured prominently in studies at the intersection or organization studies and social movements, but that have not specifically been analyzed in the context of a specific organizational network.

As for relational legacies, we refer to those pre-existing patterns of social and business relationships that have been roundly shown in the literature on "embeddedness" to shape actors understandings of what is and is not in their interest by giving them insights into the motives, trustworthiness, and capabilities of others (Granovetter 1985; Uzzi 1996, 1997). As Campbell (2005, p. 61) notes, there is a long tradition in sociology—and especially in studies of social movements—looking at networks as "mobilizing structures." As a matter of fact, these relationships affect the formation of individual interests since it is just through them that other people's perceptions, reasons, actions, and capabilities are transmitted. Ultimately, these relationships are the key constituent of social networks and serve the function of mobilizing structures. We draw on that tradition, though emphasizing also that the concept should be understood dynamically (what happens today is pre-existing for interactions that occur tomorrow), and that there is no reason to presume that all actors in a particular organizational unit will have the same relational legacies vis-à-vis actors in other organizational units. In this sense, we put particular emphasis on the concept of *path-dependence* and on the influence of the pre-existing relationships on the evolutionary network dynamics. When looking at these relational legacies, we should put apart a simplifying perspective, largely based on the univocal individuation of nodes and ties. Indeed, network relationships must be seen as "many-to-many" in nature, since they often involve multiple organizational units within the same firm and also externally the firm's boundaries, then emerging and developing at different levels and reciprocally interrelated.

In speaking of ideology (i.e., the second mobilization mechanism), the concept was initially adopted in managerial literature (Barley and Kunda 1992; Beyer et al. 1988; McKinley et al. 1998) to put into question the traditional view of theories and models of management. Indeed, contrary to depicting them as scientific, apolitical, and rational descriptions, authors in this tradition tend to put on the foreground the role played by assumptions and ideological meanings in promoting the mentioned models and theories (Parush 2008). Nevertheless, the concept of ideology adopted in this study serves the function to highlight the pathological relationship existing between managerial models and concepts such as power, authority, and control within organizations. According to this view, power is seen as a static resource, pre-assigned to well-defined individuals or groups.

Our understanding of managerial ideology follows instead a more neutral meaning, namely that suggested by Barley and Kunda (1992, p. 363), who have defined ideology as "a stream of discourse that promulgates, however unwittingly, a set of assumptions about the nature of the objects with which it deals", including the assumptions about the likely outcomes of actions under conditions of uncertainty. In studying the mobilization processes occurring in the organizational network we assume, as in Beyer et al. (1988, p. 483), that within organizations, ideologies arise not only at individual level (for instance, in the managers' mind) but, on the contrary, "can crystallize within virtually any long-lasting human group, including national cultures, social classes, professional groups, formal organizations, and organizational subunits". Accordingly, any actor within the organizational network makes assumptions about the consequences of certain actions under conditions of uncertainty; therefore, a number of coherent and identifiable ideologies can emerge and coexist within the network leading to the formation of political coalitions. In addition, when considering the power dimension, this should be conceived not as a static resource but more similarly to the way it has been promulgated by the latest studies on *management fashion*, where the emphasis is on the political strategies unfolding during daily practices and adopted by relevant actors in order to gain power (Parush 2008).

In accordance with other contributions in organizational studies (e.g., Kaplan 2008), by relying on the concept of ideology we want to overcome an analytical perspective that puts excessive emphasis on *framing* processes, seeing them as merely instrumental. Indeed, in our analysis, ideologies and framing are viewed as complementary rather than opposing concepts. The debate about the relationship between *ideology* and *frame* is alive in social movement studies (Oliver and Johnston 2000; Snow and Benford 2005; Westby 2005) where they are considered both useful to understand mobilization processes (Snow and Benford 2005). On the one hand, *frames* are "individual cognitive structures, located within the black box of mental life that which orient and guide interpretation of individual experience" (Oliver and Johnston 2000, p. 4) whereas *framing* means the use of metaphors, symbols, and cognitive cues that actors use to strategically create and manipulate "shared understandings of the world, its problems, and viable courses of action" (Campbell 2005, p. 49); on the other hand, *ideology* represents "a set of beliefs about the social world and how it operates, containing statements about the rightness of certain social arrangements and what action would be undertaken in the light of those statements. Ideology thus serves both as a clue to understanding and as a guide to action, developing in the mind of its adherents an image of the process by which desired changes can best be achieved" (Wilson 1973, p. 91).

According to Oliver and Johnston (2000), ideology focuses attention on "systems of belief, on the multiple dimensions of these belief systems, and on the ways in which the ideas are related to each other".

Although applied with regard to social movements, such an understanding of ideology is very close to that adopted by Barley and Kunda (1992), according to which it enables to interpret reality and guide action on the basis of the assumptions made on the possible results of those actions.

In our argument, similar to what has been pointed in the study of social movements, framing is not sufficient to explain the dynamic evolution of the network. This is the reason captured in Oliver and Johnston's (2000) observation that framing really "points to process, while ideology points to content". Indeed, we do not aim to simply explain just who wins political struggles, but also how the outcomes of those struggles substantively affect the dynamic evolution of the network. In other words, whereas through the concept of *framing* we can understand why certain cognitive frames "resonate" in a specific context, an understanding of the ideological contents, ideas, opinions, and meanings underlying particular collective action, when combined with attention to the effects on mobilization of relational legacies, can help us to understand the substantive nature and implications of the actual demands those winning factions and coalitions place on the executive.

In this respect, our argument is consistent with the recent strategy process literature (Kaplan 2008, p. 730), where it is argued that an analysis of the "framing contest" shows "how actors attempt to transform their own cognitive frames into the organization's predominant collective frames through their daily interactions," and "framing practices define what is at stake and thus are a means of transforming actors' interests." However, since we want to depict the network firm as dynamic and to be cognizant of the blurring of organizational boundaries, we believe that an analysis that focuses only on frames, or only on a particular focal organization, is inherently unsatisfactory. In particular, we argued that attention to ideology—understood as system of beliefs discursively maintained—can help us to understand not just how factions and coalitions form, but also the nature of the actual demands those factions and coalitions place on the executive. And we have argued that attention to relationships with parties external to the focal organization can help to understand why "framing contests" play out as they do. What we have termed relational legacies therefore shape intra- and inter-firm patterns of alliance and thus drive differences in the "stream[s] of discourse that promulgate[d], however unwittingly, [sets] of assumptions about the nature of the objects with which [they] dealt" (Barley and Kunda 1992).

6 Conclusions: A Political Perspective in Network Analysis

Organizational change and adaptation are at the core of evolutionary and behavioral theories of the firm (Cyert and March 1963; Nelson and Winter 1982). In uncertain and turbulent environments, identifying mechanisms that shape decision-making processes as well as the influence that specific action paths can potentially have on change and adaptation becomes crucial not only for theoretical development, but also for the relevant managerial implications that can be derived from such theoretical development. In this respect, the adoption of a political perspective in studying the behavior of a firm—conceived as a system of conflicts (Cyert and March 1963)—has been extensively used in the organizational and managerial literature as an interpretative lens to understand decision-making processes and related strategic and organizational implications.

The chapter has examined some central issues concerning the adoption of a political perspective in the study of network and relational dynamics. Such a perspective offers some interesting cues especially in light of the dramatic changes occurred in organizational models, increasingly based on vertical disintegration, outsourcing, and *networking*. The progressive adoption of new organizational models—documented over the past few decades in several firms and industries—has resulted in profound changes in the organizational boundaries of firm. Indeed, they became more difficult to identify in a clear and univocal manner, due to the complex, frequently ambiguous, and bilateral relationships, that developed as a consequence of remarkable changes in the organization of production and innovation activities, in the criteria of allocating tasks between the focal firm and external actors and in the governance mechanisms adopted to regulate the functioning of the network.

The increasing interdependencies that, as a consequence of these changes, emerged between internal and external, formal and informal networks, as well as the influence of their evolutionary dynamic on firm behavior easily puts on the foreground the need to adopt an interpretative model where these dynamics, the processes through which they affect firm's strategic choices and their consequences for the organization and its network, are explicitly examined through relying on a political lens.

In this chapter, we started from a critical evaluation of March's work (1962) along with the organizational politics literature that is theoretically grounded in this seminal contribution. We drew on it to debate about the relevance of depicting the network firm as a political coalition in a context where the recourse to the network is highly frequent while the organizational boundaries can be described as *blurred-but-existent*. In line with March (1964) and following studies, we recognized the central role of power and politics in shaping not only the business organization, but also and especially the network to which it belongs, as well as the existence, at any given time, of coalitions and conflicting interest systems, within groups, departments, functional units, but also across the organizational boundaries.

If we admit, as March (1962) did, that conflict resolution among cross-firm coalitions cannot be achieved through establishing a superordinate goal or leveraging simple payments and agreements, the assumption that "variation in system behavior due to conflict within the subsystem is trivial because of scale differences between the conflict within the subsystem on the one hand and conflict among subsystems on the other" (March 1962, p. 664) cannot be considered much realistic. Indeed, the growing complexity of pre-existing and emerging relationships, at different levels, both within and across the formal boundaries of the organization, constrain us to admit, more realistically, that systems of preferences could not be assumed to be well-defined and known a priori.

Drawing on more recent studies on social movements, we tried to put on the foreground the complex network dynamics that emerge and develop within and across blurring-but-existing organizational boundaries and the consequent need to consider the reciprocal influence that exists between politics, firm behavior, and network dynamics. In this attempt, we explicitly wanted to put apart a static image of the network, strongly anchored to concepts such as power, resources, and structure. On the contrary, fresh insights from the social movements literature helped promote a dynamic image of the network and, accordingly, shifted the focus onto the complex processes of coalition formation that happens across-firm as well as the role that social networks, playing as mobilizing structures, have on conflict resolution, specific demands made by coalitions, and final decisions taken by the executive.

In the study on social movements the political dimension plays a key role during the processes of social mobilization thus shaping the formation of alliances and coalitions among individuals and groups. In these processes, social networks act as mobilizing structures, conveying ideologies, opinions, and meanings and fostering or impeding the circulation of perceptions, action, and individual capabilities. These networks, that emerge and develop both within and cross-firm, shape the emergence of individual interests and it is just through networks that framing practices are put into action hence influence collecting action and foster the resonance of specific frames and ideologies among coalitions' members.

Through a critical analysis conducted across studies of organizational politics, organizational sociology, and strategic management, this work aimed at showing the advantage of adopting a political perspective to shed light on the mechanisms underlying the functioning of the network as a form of governance (with an emphasis on the evolution of both formal and informal network relationships). In particular, drawing on the social movement literature, the paper has intended to provide a precise language and a set of theoretical concepts to be used in explaining the network firm as a political coalition.

To date, efforts made in this direction by the organizational and managerial literature are still scant. In this respect, the works of Mackenzie (2008) and Whitford and Zirpoli (2009) can be viewed as an important point of departure to ground up a substantially unexplored but valuable research area within the broader field of network theory.

References

A. Arora, A. Fosfuri, A. Gambardella, *Markets for Technology: The Economics of Innovation and Corporate Strategy* (MIT Press, Cambridge, 2001)

P. Azoulay, Agents of embeddedness, NBER Working Paper 10142 (2003)

S.R. Barley, G. Kunda, Design and devotion: surges of rational and normative ideologies of control in managerial discourse. Adm. Sci. Q. **37**(3), 363–399 (1992)

C.A. Bartlett, S. Goshal, Beyond the M-form: toward a managerial theory of the firm. Strateg. Manag. J. **14**, 23–46 (1993)

R. Benford, An insider's critique of the social movement framing perspective. Sociol. Inq. **67**, 49–430 (1997)

R.D. Benford, D.A. Snow, Framing processes and social movements: an overview and assessment. Annu. Rev. Sociol. **26**, 611–639 (2000)

J. Beyer, R. Dunbar, A. Meyer, The concept of ideology in organizational analysis. Acad. Manag. Rev. **13**(1), 483–489 (1988)

J.L. Bower, *Managing the Resource Allocation Process: A Study of Corporate Planning and Investment*, 2nd edn. (Harvard Business School Press, Boston, 1970)

J.L. Bower, Y. Doz, Strategy formulation: a social and political process, in *Strategic Management: A New View of Business Policy and Planning*, ed. by D.E. Schendel, C.W. Hofer (Little Brown, Boston, 1979), pp. 152–166

R.A. Burgelman, A process model of internal corporate venturing in the diversified major firm. Adm. Sci. Q. **28**(2), 223–244 (1983)

R.A. Burgelman, Fading memories: a process theory of strategic business exit in dynamic environments. Adm. Sci. Q. **39**(1), 24–56 (1994)

J.L. Campbell, Where do we stand? Common mechanisms in organizations and social movements research, in *Social Movements and Organization Theory*, ed. by G. Davis, D. McAdam, W.R. Scott, M. Zald (Cambridge University Press, Cambridge, 2005), pp. 41–68

B. Carruthers, B. Uzzi, Economic sociology in the new millennium. Contemp. Sociol. **29**(3), 486–494 (2000)

E. Clemens, Two kinds of stuff: the current encounter of social movements and organizations, in *Social Movements and Organization Theory*, ed. by G. Davis, D. McAdam, W.R. Scott, M. Zald (Cambridge University Press, Cambridge, 2005), pp. 351–366

R. Cross, S. Borgatti, A. Parker, Making invisible work visible. Calif. Manag. Rev. **44**(2), 25–46 (2002)

R.M. Cyert, J.G. March, *A Behavioral Theory of the Firm* (Prentice-Hall, Englewood Cliffs, NJ, 1963)

G.F. Davis, T.A. Thompson, A social movement perspective on corporate control. Adm. Sci. Q. **39**(1), 141–173 (1994)

G.F. Davis, D. McAdam, Corporations, classes and social movements after managerialism. Res. Organ. Behav. **22**, 195–238 (2000)

G. Davis, M. Zald, Social change, social theory, and the convergence of movements and organizations, in *Social Movements and Organization Theory*, ed. by G. Davis, D. McAdam, W.R. Scott, M. Zald (Cambridge University Press, Cambridge, 2005), pp. 335–350

G. Davis, D. McAdam, W.R. Scott (eds.), Social movements and organization theory, (Cambridge University Press, Cambridge, 2005)

M. Diani, Introduction: social movements, contentious actions, and social networks: 'from metaphor to substance'?, in *Social Movements and Networks*, ed. by M. Diani, D. McAdam (Oxford University Press, Oxford, 2003), pp. 1–18

P.J. DiMaggio, Constructing an organizational field as a professional project: U.S. Art Museums, 1920–1940, in *The New Institutionalism in Organizational Analysis*, ed. by W.W. Powell, P.J. DiMaggio (University of Chicago Press, Chicago, 1991), pp. 267–292

P. DiMaggio, *The Twenty-First-Century Firm: Changing Economic Organization in International Perspective* (Princeton University Press, Princeton, 2001)

P.J. DiMaggio, W.W. Powell, The iron cage revisited: institutional isomorphism and collective rationality in organizational fields. Am. Sociol. Rev. **48**(2), 147–160 (1983)

J.H. Dyer, Specialized supplier networks as a source of competitive advantage: evidence from the auto industry. Strateg. Manag. J. **17**(4), 271–291 (1996)

J.H. Dyer, K. Nobeoka, Creating and managing a high-performance knowledge sharing network: the Toyota case. Strateg. Manag. J. **21**(3), 345–367 (2000)

B. Edwards, J.D. McCarthy, Resources and social movement mobilization, in *The Blackwell Companion to Social Movements*, ed. by D.A. Snow, S.A. Soule, H. Kriesi (Blackwell Publishing, Oxford, 2004)

U. Elg, U. Johansson, Decision making in inter-firm networks as a political process. Organ. Stud. **18**(3), 361–384 (1997)

N. Fligstein, I. Maria-Drita, How to make a market: reflections on the attempt to create a single market in the European Union. Am. J. Sociol. **102**(1), 1–33 (1996)

J. Gadrey, G. Gallouj, *Productivity, Innovation and Knowledge in Services: New Economic and Socio-economic Approaches* (Edward Elgar Publishing Limited, Cheltenham, 2002)

W.A. Gamson, *The Strategy of Social Protest* (Dorsey, Homewood, 1975)

E. Goffman, *Frame Analysis: An Essay on the Organization of the Experience* (Harper Colophon, New York, 1974)

G. Gavetti, D.A. Levinthal, The strategy field from the perspective of management science: divergent strands and possible integration, Management Science **50**(10), 1309–1318 (2004)

M. Granovetter, Economic action and social structure: the problem of embeddedness. Am. J. Sociol. **91**(3), 481–510 (1985)

R. Gulati, M. Gargiulo, Where do interorganizational networks come from? Am. J. Sociol. **104**(5), 1439–1493 (1999)

J. Hagedoorn, G. Duysters, Learning in dynamic inter-firm networks: the efficacy of multiple contacts. Organ. Stud. **23**(4), 525–548 (2002)

M.T. Hansen, J.M. Podolny, J. Pfeffer, So many ties, so little time: a task contingency perspective on corporate social capital in organizations, in *Social Capital of Organizations, Research in the Sociology of Organizations*, vol. 18, ed. by M. Lounsbury (Emerald Group Publishing Limited, Bingley, 2001), pp. 21–57

C. Hardy, Understanding power: bringing about strategic change. Br. J. Manag. **7**, S3–S16 (1996)

C. Hardy, N. Phillips, T.B. Lawrence, Resources, knowledge and influence: the organizational effects of interorganizational collaboration. J. Manag. Stud. **40**(2), 321–347 (2003)

G. Hedlund, A model of knowledge management and he N-form corporation. Strateg. Manag. J. **15**(S2), 73–90 (1994)

S. Helper, J.P. MacDuffie, C. Sabel, Pragmatic collaborations: advancing knowledge while controlling opportunism. Ind. Corp. Change **9**(3), 443–488 (2000)

G. Herrigel, Emerging strategies and forms of governance in the components industry in high wage regions. Ind. Innov. **11**(1/2), 45–79 (2004)

C. Jones, W. Hesterly, S. Borgatti, A general theory of network governance: exchange conditions and social mechanisms. Acad. Manag. Rev. **22**(4), 911–945 (1997)

S. Kaplan, Framing contests: strategy making under uncertainty. Organ. Sci. **19**(5), 729–752 (2008)

D. Krackhardt, Assessing the political landscape: structure, cognition, and power in organizations. Adm. Sci. Q. **35**, 342–369 (1990)

P.H. Kristensen, J. Zeitlin, *Local Players in Global Games: The Strategic Constitution of a Multinational Corporation* (Oxford University Press, Oxford, 2005)

R.N. Langlois, The vanishing hand: the changing dynamics of industrial capitalism. Ind. Corp. Change **12**(2), 297–313 (2003)

H. Leblebici, G.R. Salancik, Stability in interorganizational exchanges: rulemaking processes of the Chicago Board of Trade. Adm. Sci. Q. **27**(3), 227–242 (1982)

J. Lerner, J. Tirole, Some simple economics of open source. J. Ind. Econ. **50**(2), 197–234 (2002)

N. Levina, W. Orlikowski, Understanding shifting power relations within and across organizations: a critical genre analysis. Acad. Manag. J. **52**(4), 672–703 (2009)

G. Lorenzoni, A. Lipparini, The leveraging of interfirm relationships as a distinctive organizational capability: a longitudinal study. Strateg. Manag. J. **20**(4), 317–338 (1999)

R. MacKenzie, From networks to hierarchies: the construction of a subcontracting regime in the Irish Telecommunications Industry. Organ. Stud. **29**(6), 867–886 (2008)

J. March, The business firm as a political coalition. J. Politics **24**(4), 662–678 (1962)

J. March, H.A. Simon, *Organizations* (Wiley, New York, 1958)

D. McAdam, W.R. Scott, Organizations and movements, in *Social Movements and Organization Theory*, ed. by G. Davis, D. McAdam, W.R. Scott, M. Zald (Cambridge University Press, Cambridge, 2005), pp. 4–40

D. McAdam, J.D. McCarthy, M.N. Zald, Introduction: opportunities, mobilizing structures, and framing processes—toward a synthetic, comparative perspective on social movements, in *Comparative Perspectives on Social Movements: Political Opportunities, Mobilizing Structures, and Cultural Framings*, ed. by D. McAdam, J.D. McCarthy, M.N. Zald (Cambridge University Press, Cambridge, 1996), pp. 1–20

W. Mckinley, M.A. Mone, V.L. Barker III, Some ideological foundations of organizational downsizing. J. Manag. Inq. **7**(3), 198–212 (1998)

I. McLoughlin, C. Koch, K. Dickson, What's this "tosh"?: innovation networks and new product development as a political process. Int. J. Innov. Manag. **5**(3), 275–298 (2001)

R.E. Miles, C.C. Snow, The new network firm: a spherical structure built on a human investment philosophy. Organ. Dyn. **23**(4), 4–18 (1995)

J. Nahapiet, S. Ghoshal, Social capital, intellectual capital, and the organizational advantage. Acad. Manag. Rev. **23**(2), 242–266 (1998)

R. Nelson, S. Winter, *An Evolutionary Theory of Economic Change* (Belknap Press, Cambridge, 1982)

N. Nohria, R.G. Eccles, *Networks and Organization: Structure, Form, and Action* (Harvard Business School Press, Boston, 1992)

P. Oliver, H. Johnston, What a good idea: frames and ideologies in social movements research. Paper prepared for the Annual Meeting of the American Sociological Association, August 8, 1999, pp. 1–24 (2000)

W.J. Orlikowski, Using technology and constituting structures: a practice lens for studying technology in organizations. Organ. Sci. **11**(4), 404–428 (2000)

A. Parmigiani, W. Mitchell, Complementarity, capabilities, and the boundaries of the firm: the impact of within-firm and inter-firm expertise on concurrent sourcing of complementary components. Strateg. Manag. J. **30**(10), 1065–1091 (2009)

T. Parush, From 'management ideology' to 'management fashion': a comparative analysis of two key concepts in the sociology of management knowledge. Int. Stud. Manag. Organ. **38**(1), 48–70 (2008)

F. Passy, Social networks matter. But how?, in *Social Movements and Networks*, ed. by M. Diani, D. McAdam (Oxford University Press, Oxford, 2003), pp. 21–48

A. Pettigrew, *The Politics of Organizational Decision Making* (Tavlstock, London, 1973)

A. Pettigrew, Strategy formulation as a political process. Int. Stud. Manag. Organ. **7**, 78–87 (1977)

J. Pfeffer, *Power in Organizations* (Pitman, Marshfield, 1981)

J. Pfeffer, G. Salancik, *The External Control of Organizations* (Harper & Row, New York, 1978)

J. Podolny, J.N. Baron, Resources and relationships: social networks and mobility in the workplace. Am. Sociol. Rev. **62**(5), 673–693 (1997)

J. Podolny, K. Page, Network forms of organization. Annu. Rev. Sociol. **24**, 57–76 (1998)

W.W. Powell, Neither market nor hierarchy: network forms of organization, in *Research in Organizational Behaviour*, ed. by B. Staw, L.L. Cummings (JAI Press, Greenwich, 1990), pp. 295–336

W.W. Powell, Inter-organizational collaboration in the biotechnology industry. J. Inst. Theor. Econ. **120**(1), 197–215 (1996)

W.W. Powell, Learning from collaboration: knowledge and networks in the biotechnology and pharmaceutical industries. Calif. Manag. Rev. **40**(3), 228–240 (1998)

W. Powell, K. Koput, L. Smith-Doerr, Interorganizational collaboration and the locus of innovation: networks of learning in biotechnology. Adm. Sci. Q. **41**(1), 116–145 (1996)

H. Rao, C. Morrill, M.N. Zald, Power plays: how social movements and collective action create new organizational forms. Res. Organ. Behav. **22**, 239–282 (2000)

G. Salancik, An index of subgroup influence in dependency networks. Adm. Sci. Q. **31**(2), 194–211 (1986)

L. Smith-Doerr, W. Powell, Networks and economic life, in *The Handbook of Economic Sociology*, 2nd edn., ed. by N. Smelser, R. Swedberg (Princeton University Press, Princeton, 2005), pp. 379–402

D.A. Snow, R.D. Benford, Ideology, frame resonance, and participation mobilization. Int. Mov. Res. **1**, 197–217 (1988)

D.A. Snow, R.D. Benford, Clarifying the relationship between framing and ideology, in *Frames of Protest: Social Movements and the Framing Perspective*, ed. by H. Johnston, J.A. Noakes (Rowman & Littlefield Publishers, Inc., Maryland, 2005), pp. 205–212

D.A. Snow, E.B. Rochford Jr., S.K. Worden, R.D. Benford, Frame alignment processes, micromobilization and movement participation. Am. Sociol. Rev. **51**(4), 464–481 (1986)

T.J. Sturgeon, Modular production networks: a new American model of industrial organization. Ind. Corp. Change **11**(3), 451–496 (2002)

V.A. Thompson, *Modern Organization* (Alfred A. Knopf, New York, 1961)

C. Tilly, *From Mobilization to Revolution* (Addison-Wesley, Reading, 1978)

B. Uzzi, The sources and consequences of embeddedness for the economic performance of organizations: the network effect. Am. Sociol. Rev. **61**(4), 674–698 (1996)

B. Uzzi, Social structure and competition in interfirm networks: the paradox of embeddedness. Adm. Sci. Q. **42**(1), 35–67 (1997)

D. Westby, Strategic imperative, ideology and frames, in *Frames of Protest: Social Movements and the Framing Perspective*, ed. by H. Johnston, J.A. Noakes (Rowman & Littlefield Publishers, Inc., Maryland, 2005), pp. 217–236

J. Whitford, *The New Old Economy: Networks, Institutions, and the Organizational Transformation of American Manufacturing* (Oxford University Press, Oxford, 2005)

Whitford, J., Zirpoli, F. (2009) The (vertical) network firm as a political coalition: the reorganization of Fiat Auto, Maggio, SSRN: http://ssrn.com/abstract=1426860

R.H.Williams, R.D. Benford, Two faces of collective action frames: a theoretical consideration, paper presented at the annual meeting of the Midwest Sociological Society, Chicago (1996)

J. Wilson, *Introduction to Social Movements* (Basic Books, New York, 1973)

M.N. Zald, M.A. Berger, Social movements in organizations coup d'etat, insurgency, and mass movement. Am. J. Sociol. **83**(4), 823–861 (1978)

Chapter 11
Markets of Logistics Services: The Role of Actors' Behavior to Enhance Performance

Nicola Bellantuono, Gregory E. Kersten and Pierpaolo Pontrandolfo

Abstract In real markets of logistics services, actors make independent decisions to pursue their own objectives, neglecting the need for maximizing performance of the market as a whole. The aim of this chapter is to assess the inefficiency of such logistics markets and define policies to improve system-wide performance, taking into account each actor's behavior. A simulation model of a logistics marketplace is thus defined, wherein the transportation needs of a number of shippers have to be matched with the capacities of several carriers. The model is used to assess the players' behavior and system performance in a decentralized logistics market. The analysis shows the extent to which certain features of the market affect inefficiency, stressing the room for improvement. Based on simulation results, several recommendations are given, aimed at influencing the actors' autonomous decision making. We discuss how the recommendations' efficacy is impacted by behavioral issues.

N. Bellantuono (✉) · P. Pontrandolfo
Department of Mechanics, Mathematics and Management, Polytechnic University of Bari,
Viale Japigia, 182 70126 Bari, Italy
e-mail: n.bellantuono@poliba.it

P. Pontrandolfo
e-mail: pontrandolfo@poliba.it

G. E. Kersten
InterNeg Research Centre, Concordia University, 1450 Guy Street Montreal,
Quebec H3H 0A1, Canada
e-mail: gregory@jmsb.concordia.ca

I. Giannoccaro (ed.), *Behavioral Issues in Operations Management*,
DOI: 10.1007/978-1-4471-4878-4_11, © Springer-Verlag London 2013

1 Introduction

In supply chains (SCs) managed in a decentralized fashion, actors autonomously make decisions by defining the logistics policies (mostly dealing with inventory management) that maximize their own utilities, regardless of the system-wide efficiency. Research has shown that decentralized SCs prove inefficiently (Cachon and Zipkin 1999; Cooper et al. 1997; Frohlich and Westbrook 2001; Vickery et al 2003). Several real SCs are not centrally managed, and in particular, often, logistics services are exchanged through pure markets, wherein decisions are made under a totally decentralized fashion (Ağralı et al. 2008).

Centralized management, which is consistent with the optimization of the SC as a whole, is based on hypotheses that are barely realistic. It postulates the existence of an actor (below also referred to as decision maker) who: (1) owns all the relevant information along the chain, (2) is able to define policies, which are optimal under a system-wide perspective, and (3) has the bargaining power to make the other actors behave in accordance with such policies.

In reality, actors usually have little access to inform about other SC stages (Corbett and Tang 1999) and are affected by bounded rationality (Simon 1982; Rubinstein 1998), which prevents them from identifying a true globally optimal policy (Su 2008). Furthermore, the opportunistic behavior of all the parties makes it difficult to put into practice the optimal global policy, even if identified (Lee and Whang 1999; Nyaga et al. 2010).

Thus, it is very common that SCs operate in a decentralized fashion: all actors act as decision makers and, based on partial information, define and adopt policies that they consider effective for the SC stage.

In logistics and transportation, decentralization is more frequent than in manufacturing processes. Recently, manufacturers increasingly entrust the logistics functions of their operations to third-party logistics providers (3PLs), who provide one or more specialized services on behalf of their customers. The variety of logistics services and specialization of their providers, coupled with the need for a higher integration in the supply chain, resulted in the appearance of fourth-party logistics providers (4PLs), i.e., integrators capable of delivering complete solutions, from the strategic design of the logistics network to the day-by-day operational issues (Yao 2010). The emergence of 3PLs and 4PLs has determined a growing coordination among the different logistics services. However this is not enough to guarantee an adequate SC coordination, in particular at the interface between logistics providers and manufacturing companies, or, generally speaking, between carriers and shippers, where a shipper is either a manufacturing company or a company that demands logistics services on behalf of third parties.

Specific additional problems of logistics are associated with the nature of services. They are indeed intangible, heterogeneous, perishable, their production is inseparable from their consumption (Zeithaml et al. 1985), and often requires customization. All such features usually make it difficult to measure specifications and performances of services (Fitzsimmons et al. 1998). In particular, logistics services are affected by additional sources of complexity and widely range from basic to advanced services (Andersson and Norrman 2002; Giannoccaro et al. 2009). Moreover, services in some cases are procured as a bundle (Schoenherr and Mabert 2008). This results in the need to consider several attributes beside price for managing logistics services procurement, i.e., lead time, time flexibility, occurrence of delays and associated penalties, etc.

Based on the above assumption that centralized decision making is neither common in reality nor easy to be implemented, this paper discusses ways to enhance SC coordination, with emphasis on logistics markets, in the context of decentralized decision making processes. We propose the concept of the "organized market of logistics services". This concept is mainly based on managing information which the actors receive in order to counterbalance lack of a single centralized decision maker. A possible way to achieve this is the adoption of schemes such as supply contracts (Tsay et al. 1999; Tang 2006), which are mechanisms based on incentives coordinating transactions among two or more SC actors.

As the proposed approach is based on information management and leverages on autonomous decisions (and related actions) by the SC actors, it turns out that behavioral issues are the key. In fact, once decisions and actions are identified and coherent incentive schemes designed, the actual implementation of actions relies on such issues as understanding of the rules to share costs, benefits, and risk, reciprocal trust (Cummings and Bromiley 1996; Zaheer et al. 1998) and perception of opportunism (Williamson 1985; Rousseau et al. 1998; Brown et al. 2000).

The specific aims of this paper are (1) to assess the inefficiency associated with decentralized decision making approach in a market of logistics services and (2) to suggest recommendations to guide actors in their decision making in order to "organize the market". Such actions leverage on a suitable information management, based on the assumption that more effective decisions stem from better information management.

With respect to the first objective, we define a stylized simulation model of the market of logistics services as a marketplace wherein the transportation needs of a number of shippers can be matched with the capacities of several carriers. On the supply side, each carrier is characterized by capacity, cost structure, and pricing strategy. On the demand side, each request for transportation (which is posted by a shipper) is defined by the quantity to be moved and the route (in turn specified by the points of origin and destination). All the shipments are assumed to be made at the same time.

We use simulation in order to determine the allocation of transportation requests to carriers. Because the actors make autonomous decisions, the allocation solution is likely to be inefficient, i.e., there are other solutions with lower total transportation costs for shippers and/or higher margins for carriers.

To assess inefficiency, we developed two heuristics. The first heuristic decisions are based on the minimization of every price a shipper pays to the carriers and the maximization of each carrier margin. In the second heuristic, the criterion of these actors describes global rather than local prices and margins. That is every shipper and carrier select solution that lower total costs. Note that the second heuristic moves the process from local toward global optimization.

Based on the simulation results, we assessed the potential for performance improvement in markets of logistics services, prepared guidelines to pursue such an improvement in real systems characterized by decentralized decision making, and argued for the need for research on the behavioral issues related to logistics procurement.

The paper is organized as follows. Section 2 introduces the simulation model of the logistics market, discusses the two heuristics utilized to simulate the (decentralized) decision-making process by which transportation requests are assigned to carriers, and presents performance measures to evaluate the solution to the transportation problem. In Sect. 3, we discuss simulation results, which lead us to propose recommendations in Sect. 4 regarding the organization of logistics markets. In Sect. 5, we point out some limitations of our study, derive the main managerial implications, and suggest avenues for future research.

2 The Simulation Model

We define logistics market as a marketplace wherein several companies interact to provide or acquire logistics services. Modeling a real-like logistics market is not trivial due to the number and the variety of involved actors, the complex relationships among them, and the features of the logistics service, which, in turn, require exchange of information other than the price to complement the offer; for instance, information may include technical service specifications such as quality and load assurance and special price schemes that rule delays, rush rates, and early reservations.

The logistics market may also be modeled in a simplified way. We consider two different sets of actors: m shippers ($i = 1, ..., m$), each requesting one transportation service, and n carriers ($j = 1, ..., n$) who are able to provide those services. The problem lies in allocating each transportation request to a carrier. Carriers are heterogeneous in terms of cost structures and price strategies, and have a finite capacity K_j.

Each transportation consists in moving a quantity q_i of a good along its route, namely from an origin to a destination, and repositioning the vehicle in the point of origin once the delivery has been completed. The length of the i-th transportation route is calculated as Euclidean distance:

$$d_i = \sqrt{(x_{Ai} - x_{Bi})^2 + (y_{Ai} - y_{Bi})^2}, \tag{1}$$

where in (x_{Ai}, y_{Ai}) and (x_{Bi}, y_{Bi}) are the coordinates of the origin and the destination, respectively.

We assume that all the shipments occur at the same time, therefore, each vehicle can be used along one path only. However, in order to account for the possible transportation of goods on vehicles returning to their points of origin we allow their use in both directions.

For the carriers' cost structures, the transportation cost (c_{ij}) sustained by the j-th carrier for the transportation service i depends on his per-mileage cost per unit of shipment (u_j) and is affected by two forms of economies of scale, which make the transportation cost increase in distance (i.e., path length) and in quantity less than proportionally. The existence of fixed costs per shipments (e.g., loading and unloading costs) explains the occurrence of the economies of scale associated with distance, whereas a cost per payload, which is lesser for large vehicles than for small ones, results in economies of scale associated with quantity. Both kinds of economies of scale follow a power function, with exponents α_j for distance and β_j for quantity ($0 < \alpha_j \leq 1$; $0 < \beta_j \leq 1$).

Thus the cost that the j-th carrier would sustain to provide the i-th service is:

$$c_{ij} = u_j d_i^{\alpha_j} q_i^{\beta_j}. \qquad (2)$$

The cost of providing the service may be affected by savings resulting from putting together transportation requests and allocating them to the same carrier as well as from the similarities of the requests. Similarities mostly are associated with transportation optimization. Although this issue is out of scope, we note here about two forms of similarities in transportation:

- *similarity for consolidation* occurs when two or more shipments along the same route (i.e., with the same origin and the same destination) are provided by the same carrier; the carrier benefits from the economies of scale are associated with quantity;
- *similarity for repositioning* occurs when two or more shipments along opposite routes (i.e., whose origins and destinations are reversed) are provided by the same carrier; the carrier avoids the cost of repositioning the vehicle to the origin after its usage.

It is assumed that carriers adopt a price strategy based on a mark-up policy. In default of similarities, the price p_{ij} that the carrier j asks for the transportation i is equal to the corresponding cost, increased by the mark-up factor $\gamma_j > 0$, which is peculiar to each carrier j:

$$p_{ij} = (1 + \gamma_j)c_{ij}. \qquad (3)$$

If any similarity exists, the carriers can exploit it and reduce the price they offer by a discount factor:

$$p_{ij} = (\gamma_j + \delta_j)u_j d_i^{\alpha_j} q_i^{\beta_j} + (1 - \delta_j)c_{ij}, \qquad (4)$$

where $0 \leq \delta_j \leq 1$.

If $\delta_j = 1$, then the carrier receives all the savings because of the transportation similarity (see above). If however, $\delta_j = 0$, then the price is such that the carrier gets the same margin that he would achieve in the absence of similarities. In this case the shipper gets all the savings.

2.1 Heuristics

To match the shippers' transportation needs with the carriers' transportation capacity, two heuristics are proposed. Heuristics 1 emulates the decision-making behavior adopted in a marketplace, wherein actors do not collaborate to optimize the system-wide performance and do not share all information. This heuristics aims at providing a realistic solution in which each actor tends to pursue his or her own goals; specifically, the carriers' goal is maximizing the margin and the shippers' goal is minimizing the cost. Heuristics 2, used as a benchmark, is designed assuming that actors collaborate by sharing information so as to increase the system-wide efficiency; the latter is measured in terms of the sum of carriers' costs, that is the costs sustained by the system as a whole, assumed as a black box. In both heuristics, the actors choose their counterpart through a sequential approach.

Note that none of the two heuristics is useful in itself; instead they are used here with the purpose of assessing the inefficiency of logistics markets in which both carriers and shippers make decisions independently and with no consideration of any form of collaboration. In other words, the heuristics do not intend to provide a near-optimal solution but rather attempt to: (1) emulate real behaviors, (2) identify inefficiencies, and (3) indicate potential improvements.

2.1.1 Heuristics 1

This heuristics refer to the case of lack of collaboration among the actors and consists of five steps:

1. *Request for quotation (RfQ).* In this step, each shipper, whose transportation request needs to be allocated, issues RfQ communicating the service details (quantity, origin, and destination) to all carriers.
2. *Bidding.* All carriers, whose capacity has not yet been allocated, calculate the costs that they would sustain for each transportation request. The cost c_{ij} is primarily based on the quantity to be shipped, the transportation distance, and the carrier's cost structure. Specifically, the carrier takes into account possible similarities between each request and the transportations that they have confirmed in previous steps, if any. The cost is computed as follows:

$$c_{ij} = u_j d_i^{\alpha_j} \left[\max\left(q_i + QC_{ij}; QR_{ij}\right)^{\beta_j} - \max\left(QC_{ij}; QR_{ij}\right)^{\beta_j} \right], \quad (5)$$

where QC_{ij} and QR_{ij} denote the quantities similar for consolidation and for repositioning, respectively. Note that for $QC_{ij} = QR_{ij} = 0$, Eq. (2) holds.

Once the cost of the transportation request has been calculated, the carriers define the price through Eq. (4) and post their offers to the shippers.

3. *Offers selection.* Each shipper evaluates all the offers received by the carriers and selects the one at the lowest price. Then, she reserves certain capacity of the carrier.

4. *Reservations acknowledgment.* Carriers who have received at least one reservation calculate again the costs by taking into account similarities both with the requests confirmed at the previous steps and with the reservations received at this step, the latter being denoted as SC_{ij} and SR_{ij}:

$$c_{ij} = u_j d_i^{\alpha_j} \left[\max\left(q_i + QC_{ij} + SC_{ij}; QR_{ij} + SR_{ij}\right)^{\beta_j} - \max\left(QC_{ij}, QR_{ij}\right)^{\beta_j}; 0 \right] \cdot$$
$$\left[\frac{\max\left(q_i + SC_{ij} - SR_{ij}; 0\right) + \frac{1}{2}\min\left(q_i + SC_{ij}; SR_{ij}\right)}{\max\left(q_i + SC_{ij}; SR_{ij}\right)} \right] \frac{q_i}{q_i + SC_{ij}}. \qquad (6)$$

Note that for $SC_{ij} = SR_{ij} = 0$, Eq. (5) holds.

Since all shippers emit their reservation concurrently, it may happen that a carrier receives requests exceeding his transportation capacity. Therefore, carriers may select which reservations to confirm: to do so, they compute the margin of each reservation and confirm the reservations in a descending order of margin, until their capacity is completely allocated. Exceeding reservations, if any, are rejected.

5. *Iteration.* Steps 1–4 are repeated until all the requests are allocated to a carrier or all the carriers use up their transportation capacity.

2.1.2 Heuristics 2

In this heuristics, a mutual collaboration among actors exists, and decisions aim at minimizing the system-wide costs. This heuristics include 5 steps: steps 1, 2, and 5 are the same as in Heuristics 1, while the third and the fourth steps differ as described next:

1. *Request for quotation (RfQ).* The same as Heuristics 1.
2. *Bidding.* The same as Heuristics 1.
3. *Offers selection.* Each shipper evaluates the offers received by the carriers and selects the offer at the lowest cost (for Heuristics 1 the selection criterion is the price). Then, she reserves certain capacity of the carrier.
4. *Reservation acknowledgment.* Carriers who have received at least one reservation calculate again the costs to provide the services by using Eq. (6). Then, they compute the margin of each reservation and confirm the reservations according to an ascending order of cost, until their transportation capacity is completely allocated (for Heuristics 1 the criterion that carriers adopt to

confirm reservations is the margin). Exceeding? Reservations (exceeding something?), if any, are rejected.

5. *Iteration.* The same as Heuristics 1.

2.2 Performance Measures

Three performance measures intended to assess performance of the shippers, the carriers, and the system as a whole are defined below.

Given that the goal of each shipper is to find a carrier that provides the transportation service at the lowest price, the aggregate shippers' performance is measured as follows:

$$P = \sum_{i=1}^{m} \sum_{j=1}^{n} \bar{p}_{ij}, \tag{7}$$

where, if the service i is provided by the carrier j, $\bar{p}_{ij}$ is the price at which the reservation is confirmed; otherwise, $\bar{p}_{ij} = 0$.

The goal of carriers is to maximize their margin (calculated as the difference between price and cost). Therefore, the aggregate carriers' performance is defined as:

$$Y = \sum_{j=1}^{n} \sum_{i=1}^{m} \left(\bar{p}_{ij} - \bar{c}_{ij} \right), \tag{8}$$

where, if carrier j provides the transportation i, $\bar{p}_{ij}$ and $\bar{c}_{ij}$ are the prices at which carrier j confirms that the reservation and the cost he sustains to provide it; otherwise, $\bar{p}_{ij} = \bar{c}_{ij} = 0$.

In a system-wide perspective, the goal is to satisfy all the transportation requests in the most efficient way, i.e., at the minimum cost. Therefore, the system-wide performance is:

$$C = \sum_{j=1}^{n} \sum_{i=1}^{m} \bar{c}_{ij} = Y - P. \tag{9}$$

The performances of both heuristics are compared by adopting properly designed competition penalties. As for the system-wide performance, we use:

$$CP_C = \frac{C_1 - C_2}{C_2}, \tag{10}$$

where the indexes C_1 and C_2 refer to Heuristics 1 and 2, respectively. Similarly, the competition penalties used to compare the shippers' and the carriers' performance, are defined as follows:

$$CP_P = \frac{P_1 - P_2}{P_2}, \tag{11}$$

$$CP_Y = \frac{Y_1 - Y_2}{Y_2}. \tag{12}$$

If $CP_C > 0$, then $C_1 > C_2$. Thus Heuristics 1 underperforms Heuristics 2 in the system-wide performance. Similarly, if $CP_P > 0$, then $P_1 > P_2$, thus Heuristics 1 underperforms Heuristics 2 in the shippers' performance. Conversely, $CP_Y > 0$ means that $Y_1 > Y_2$, i.e., Heuristics 1 overperforms Heuristics 2 in the carriers' performance.

3 Simulation Results

A numerical analysis is provided to assess the inefficiency of Heuristics 1 in several scenarios. In all of them we assume $m = 100$ shippers, each requesting one transportation. For all transportation requests the quantity to be moved is equal to one, whereas the route may differ in terms of paths and direction. M paths are generated by drawing at random a couple of points in a 100×100 square. Then, to each transportation request we assign (1) a specific path by drawing at random from M paths, and (2) the path direction. The higher the M value, the higher the probability that the transportation requests are dissimilar.

A number of carriers n are available to satisfy the transportation requests. Each carrier is characterized by the per-mileage cost per unit of shipment u_j, the parameters governing the intensity of the economies of scale (α_j and β_j), the mark-up factor γ_j, and the price discount factor δ_j. The values of u_j, α_j, β_j, and γ_j are randomly assigned according to a normal distribution, while δ_j is deterministic and equal to δ for all carriers (Table 1).

A real logistics market characterized by the existence of a few large logistics providers that serve many different routes corresponds to the scenarios characterized by a low n and a high M. The real cases of a logistics market where the shippers' demand is satisfied by a high number of owner drivers is modeled by $n = 10$. The scenarios with asymmetric distribution of the capacity of the carriers resemble the logistics markets characterized by the presence of both large and small logistics providers.

In each scenario, results represent the average of 1000 replications. Table 2 shows results for the scenarios in which the capacity is equally distributed among the carriers. Table 3 illustrates/provides finding obtained from asymmetric distribution of the capacity.

As we expected, in all scenarios Heuristics 1 provides lower system-wide performance than Heuristics 2 ($CP_C > 0$). Moreover, in all scenarios both CP_P and CP_Y are positive, which means that Heuristics 2 determine lower aggregate performance of the carriers and higher aggregate performance of the shippers than

Table 1 Values of the variables characterizing the carriers

Variable	Distribution	Mean	Standard deviation
u_j	Normal	100	20
α_j	Normal	0.70	0.10
β_j	Normal	0.09	0.03
γ_j	Normal	2.00	0.20; 0.50
δ_j	Deterministic	0.30; 0.70	–

Heuristics 1. Thus, the system as a whole benefits from the collaboration among the actors, but the resulting benefits are gained solely by the shippers. For this reason, the carriers have no interest in adopting a collaborative behavior aimed at optimizing the system-wide performance. To motivate the carriers to collaborate, a contract with a clause, specifying their share of benefits obtained from collaboration, could be prepared.

Even in the case of asymmetric distribution of the capacity, Heuristics 1 underperforms Heuristics 2 at the expense of the shippers' performance. As M increases, the CP_C value rises.

Figure 1 shows how the value of CP_C is affected by both n and M given the values of δ and $\sigma(\gamma)$. In particular, as M increases, the system-wide inefficiency of Heuristics 1 strongly rises in all the cases (Fig. 1a–d) and it increases in n in most of cases. A Student's t test confirms that all differences are statistically significant ($\alpha = 0.05$). Moreover, findings show that CP_C is not affected by δ and $\sigma(\gamma)$. In fact, the Student's t test indicates that the null hypothesis cannot be rejected with $\alpha = 0.05$.

To assess the influence of the asymmetry in capacity distribution among carriers, the scenarios in which $n = 7$ with the capacity equally distributed are compared with those in which $n = 7$ with the capacity asymmetrically shared.

Table 4 reports the percent increases in CP_C moving from the symmetric (S) to asymmetric (A) capacity distribution. The higher the percent value, the higher the inefficiency of Heuristics 1 in the asymmetric case compared with the symmetric case. Data show that if the similarity among paths is high ($M = 10$ or 30), the relative inefficiency of Heuristics 1 is on average higher with an asymmetric distribution of the capacity among carriers, whereas if the transportation paths are different ($M = 100$), it is higher when the capacity is equally shared.

4 Organizing the Market

As mentioned in the introduction, in most real cases the procurement of transportation services occurs through the market as a coordination mechanism. Recent trends—specialization of logistics providers and, in particular, the emergence of integrators, 3PLs, and 4PLs, respectively—seem not enough to exploit the opportunities to improve transportation service while reducing the related costs: the performance of transportation is indeed still poor (Gorick 2006; Ergun et al. 2007).

Table 2 Simulation results for the scenarios with symmetric distribution of the capacity among carriers

Scenario				Performance		
n	M	δ	$\sigma(\gamma)$	CP_C	CP_Y	CP_P
4	10	0.3	0.2	0.0535	0.2064	0.1718
4	10	0.3	0.5	0.0724	0.2246	0.1901
4	10	0.7	0.2	0.1161	0.2537	0.2242
4	10	0.7	0.5	0.0964	0.2307	0.2020
4	30	0.3	0.2	0.1601	0.2610	0.2354
4	30	0.3	0.5	0.1351	0.2249	0.2022
4	30	0.7	0.2	0.1497	0.2448	0.2216
4	30	0.7	0.5	0.1258	0.2248	0.2005
4	100	0.3	0.2	0.2187	0.2031	0.2074
4	100	0.3	0.5	0.2384	0.2160	0.2223
4	100	0.7	0.2	0.2272	0.2157	0.2188
4	100	0.7	0.5	0.2048	0.2039	0.2041
7	10	0.3	0.2	0.0942	0.2281	0.1975
7	10	0.3	0.5	0.0970	0.2196	0.1915
7	10	0.7	0.2	0.0992	0.2081	0.1846
7	10	0.7	0.5	0.1120	0.2157	0.1932
7	30	0.3	0.2	0.1355	0.2077	0.1895
7	30	0.3	0.5	0.1289	0.1983	0.1807
7	30	0.7	0.2	0.1430	0.2044	0.1895
7	30	0.7	0.5	0.1706	0.2240	0.2111
7	100	0.3	0.2	0.2382	0.2069	0.2156
7	100	0.3	0.5	0.2438	0.2024	0.2140
7	100	0.7	0.2	0.2396	0.2139	0.2209
7	100	0.7	0.5	0.2297	0.1993	0.2076
10	10	0.3	0.2	0.1143	0.1951	0.1764
10	10	0.3	0.5	0.1190	0.1843	0.1692
10	10	0.7	0.2	0.1434	0.2071	0.1932
10	10	0.7	0.5	0.1379	0.1935	0.1813
10	30	0.3	0.2	0.1587	0.2135	0.1996
10	30	0.3	0.5	0.1559	0.1999	0.1887
10	30	0.7	0.2	0.1799	0.2085	0.2016
10	30	0.7	0.5	0.1809	0.1901	0.1879
10	100	0.3	0.2	0.2516	0.2168	0.2265
10	100	0.3	0.5	0.2405	0.1932	0.2064
10	100	0.7	0.2	0.2262	0.1933	0.2023
10	100	0.7	0.5	0.2259	0.1780	0.1911

The adoption of a coordination mechanism different from the market and characterized by a higher centralization of decision making (e.g., hierarchy) is barely realistic, due to the fragmentation of the transportation sector and the lack of a clear process owner. Acknowledging this, we claim that some improvement could be achieved by "organizing the market". To this aim, we have introduced a

Table 3 Simulation results for the scenarios with asymmetric distribution of the capacity among carriers

Scenario				Performance		
n	M	δ	$\sigma(\gamma)$	CP_C	CP_Y	CP_P
7	10	0.3	0.2	0.1022	0.2062	0.1828
7	10	0.3	0.5	0.1173	0.2196	0.1966
7	10	0.7	0.2	0.1268	0.2232	0.2025
7	10	0.7	0.5	0.1272	0.2292	0.2072
7	30	0.3	0.2	0.1560	0.2185	0.2028
7	30	0.3	0.5	0.1606	0.2208	0.2057
7	30	0.7	0.2	0.1951	0.2361	0.2262
7	30	0.7	0.5	0.1519	0.1853	0.1773
7	100	0.3	0.2	0.2429	0.2099	0.2191
7	100	0.3	0.5	0.2136	0.1826	0.1913
7	100	0.7	0.2	0.2223	0.1950	0.2024
7	100	0.7	0.5	0.2366	0.1894	0.2022

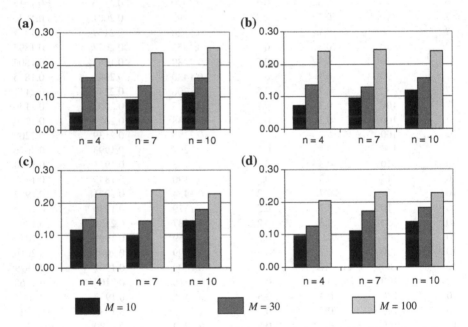

Fig. 1 CP_C values for **a** $\delta = 0.3$ and $\sigma(\gamma) = 0.2$; **b** $\delta = 0.3$ and $\sigma(\gamma) = 0.5$; **c** $\delta = 0.7$ and $\sigma(\gamma) = 0.2$; **d** $\delta = 0.7$ and $\sigma(\gamma) = 0.5$

simulation model (Sect. 2) and assessed the potential for such improvements under different scenarios and different ways of organizing logistics markets.

It is crucial then to specify what "organizing the market" is, and how this can be actually implemented (and in turn reflected in the proposed model). Rather than modifying the organizational structure, which, to sum up, concerns the design of

Table 4 Percent increase in CP_C moving from the symmetric (S) to asymmetric (A) capacity distribution

	$\delta = 0.3$		$\delta = 0.5$		mean (%)
	$\sigma(\gamma) = 0.2$ (%)	$\sigma(\gamma) = 0.5$ (%)	$\sigma(\gamma) = 0.2$ (%)	$\sigma(\gamma) = 0.5$ (%)	
$M = 10$	8.49	20.93	27.82	13.57	17.70
$M = 30$	15.13	24.59	36.43	−10.96	16.30
$M = 100$	1.97	−12.39	−7.22	3.00	−3.66

Note the percent increase is calculated as $[CP_C^{(A)} - CP_C(S)]/CP_C(S)$

the allocation of decision rights and the related communication links, we propose influencing the decision-making process and give three recommendations.

The recommendations aim at improving the match between supply and demand in logistics markets. The information exchange and the actors' decisions should be improved, so as to diminish the information processing effort and, at the same time, increase the decision effectiveness of both actor and the system as a whole. We propose the following:

1. *Select.* Help actors to focus on data or potential transaction, which are more relevant (and might be overlooked due to excess of information coupled with bounded rationality);
2. *Enrich.* Complement the data formally exchanged in the potential transactions with other relevant information;
3. *Modify.* Change the terms of the potential transactions to make them more efficient and beneficial to both parties.

In the following we provide examples for each recommendation and relate them to the model proposed in Sect. 2.

With respect to recommendation 1, it is possible to select information by: (1) identifying similar transportation requests and push carriers or shippers to jointly examine them, when formulating or selecting the offers; and (2) grouping requests consistent with offers (e.g., requests characterized by low quantity uncertainty that fit with transportation contracts involving advanced reservation).

With respect to recommendation 2, information can be enriched by complementing the transportation price with data such as the existence of some flexibility on the pick-up or delivery dates.

With respect to recommendation 3, modifying information means changing the terms of the potential transactions (e.g., contract clauses) to make certain transactions preferable compared to others, due to their higher system utility, while at the same time making sure that neither party decreases utility, should he/she agrees on such transactions.

To explain the value of "organizing" activities mentioned above, we may consider an actor (e.g., an intermediating agency) responsible for these activities. This actor uses economic incentives in order to direct both carriers and shippers. The actor's recommendations lead to increase of the system's efficiency. The incentives include transferring a portion of the savings achieved from the increased efficiency to the shippers and carriers. The incentive scheme suggested above takes into account that decisions cannot be forced on SC actors. The scheme, therefore, conforms to the following two principles (Tsay et al. 1999; Bahinipati et al. 2009):

1. *Channel coordination*, i.e., system performance has to increase compared with the "not-organized" market;
2. *Win–win condition*, i.e., every actor has to be convinced that the organization does not disadvantage her in any way.

Principle 1 ensures that there is the possibility to promote behaviors by SC actors, which are virtuous under a system-wide perspective. In fact, the whole system performance improvement means that the proverbial pie is available to be shared by the participants in the logistics market. Principle 2 deals with the criteria determining how to cut that pie to the satisfaction of every participant.

We believe that conforming to these principles requires that both rational and behavioral issues are taken into account. Decision-making problems are usually modeled based on variables related to rationality, such as availability of perfect information, nature of information (e.g., private versus public), utility functions (e.g., the risk aversion of decision makers). Such variables are relatively easy to be dealt with, however, they are likely to be not adequate to describe real problems and to identify actual solutions. They need to be complemented by other variables, those that can model behaviors. Models should take into account aspects such as trust, perception of opportunism by the counterpart, expectation for building a relationship, etc. especially for systems within which decisions are made by several independent actors, who own information related to each respective local environment. What one might expect is based only on rational variables which may be contradicted due to the influence of behavioral variables.

5 Conclusions

Most of real logistics services are exchanged through pure market mechanisms where decisions are made in a totally decentralized fashion. For instance, in transportation markets, carriers, and shippers make independent decisions in order to pursue their own objectives, neglecting the need for maximizing the performance of the market as a whole. This is mostly due to the lack of conditions that would allow decisions to be centralized and results in significant inefficiencies for the market as a whole.

To assess such inefficiencies we conducted a simulation study on a simplified model of a marketplace, wherein shippers and carriers interact to provide or acquire logistics services. We developed a heuristics simulating the players' behaviors and compared performance against a benchmark. Results show that, (1) inefficiency increases with the diversity in the routes, the number of carriers, and the asymmetric distribution of the capacity of the carriers (the negative impact of the latter is emphasized for low diversity among routes); (2) the system-wide performance seems not to be affected by (1) a carrier pricing strategy aimed at pursuing consolidation (through discounts for similar routes) and (2) a higher variance in the mark-up factor used by carriers to set prices.

To achieve performance improvements of logistics markets, one should acknowledge the allocation of decision rights as given in reality. In particular, when the decision-making process is decentralized, global optimization models are likely to prove ineffective. In this paper we suggested to "organize the market", i.e., to influence the decision-making process, in order to enhance coordination among the SC actors. To this aim we give recommendations aimed at increasing the decision-making process' effectiveness. The recommendations are of three types: *select*, *enrich*, and *modify* information exchanged by actors. It is worthwhile to stress that such recommendations conform to channel coordination and win–win conditions, which jointly assure that a potential improvement exists and every party can benefit from it.

We argued that to effectively organize the market, in addition to the variables related to rationality, behavioral issues should be taken into account. They are indeed critical in contexts where several actors interact and make independent decisions. Behavioral issues relate to subjective social perceptions and expectations regarding oneself, the counterpart, and the context in which the transaction occurs, and are affected by the actors' bounded rationality.

We believe that the adoption of the proposed recommendations can positively impact the performance of logistics markets, by actually achieving the potential improvement assessed through the proposed simulation study. Subsequent studies will attempt to identify concrete actions required for the recommendations' implementation as well as analyze the impact of behavioral aspects on the implementation.

Acknowledgments This work has been supported by Regione Puglia (APQ PS025) and the Engineering Faculty in Taranto of the Politecnico di Bari.

References

S. Ağralı, B. Tan, F. Karaesmen, Modeling and analysis of an auction-based logistics market. Eur. J. Oper. Res. **191**, 272–294 (2008)

D. Andersson, A. Norrman, Procurement of logistics services—a minutes work or a multi-year project. Eur. J. Purch. Supply Manag. **8**, 3–14 (2002)

B.K. Bahinipati, A. Kanda, S.G. Deshmukh, Coordinated supply management: review, insights, and limitations. Int. J. Logist. Res. Appl. **12**(6), 407–422 (2009)

J.R. Brown, C.S. Dev, D.J. Lee, Managing marketing channel opportunism: The efficacy of alternative governance mechanisms. J. Mark. **64**(2), 51–65 (2000)

G.P. Cachon, P.H. Zipkin, Competitive and cooperative inventory policies in a two-stage supply chain. Manage. Sci. **45**(7), 936–953 (1999)

M.C. Cooper, D.M. Lambert, J.D. Pagh, Supply chain management: more than a new name for logistics. Int. J. Logist. Manag. **8**(2), 1–14 (1997)

C.J. Corbett, C.S. Tang, Designing supply contracts: contract type and information asymmetry, in *Quantitative models for supply chain management*, ed. by S. Tayur, R. Ganeshan, M. Magazine (Kluwer Academic Publishers, London, 1999), pp. 269–297

L.L. Cummings, P. Bromiley, The organizational trust inventory (OTI): development and validation, in *Trust in Organizations: Frontiers of Theory and Research*, ed. by R.M. Kramer, T. Tyler (Sage, Thousand Oaks, 1996), pp. 302–330

O. Ergun, G. Kuyzu, M. Salvelsbergh, Reducing truckload transportation costs through collaboration. Transportation Science **41**(2), 206–221 (2007)

J.A. Fitzsimmons, J. Noh, E. Thies, Purchasing business services. J. Bus. Ind. Mark. **13**(4–5), 370–380 (1998)

M.T. Frohlich, R. Westbrook, Arcs of integration: an international study of supply chain strategies. J. Oper. Manag. **19**, 185–210 (2001)

I. Giannoccaro, R. Moramarco, P. Pontrandolfo, E-procurement of logistics services: the impact of the service attributes on the exchange mechanism. In: *20ᵗʰ International Conference on Production Research*, Shanghai, 2–6 August 2009

J. Gorick, Running on empty? Logist. Transp. Focus **8**(10), 25–26 (2006)

H. Lee, S. Whang, Decentralized multi-echelon supply chains: incentives and information. Manage. Sci. **45**(5), 633–639 (1999)

G.N. Nyaga, J.M. Whipple, D.F. Lynch, Examining supply chain relationships: Do buyer and supplier perspectives on collaborative relationships differ? J. Oper. Manag. **28**(2), 101–114 (2010)

D.M. Rousseau, S.B. Sitkin, R.S. Burt, C. Camerer, Not so different after all: A cross-discipline view of trust. Acad. Manag. Rev. **23**(7), 393–404 (1998)

A. Rubinstein, *Modeling bounded rationality* (The MIT Press, Cambridge, 1998)

T. Schoenherr, V.A. Mabert, The use of bundling in B2B online reverse auctions. J. Oper. Manag. **26**(1), 81–95 (2008)

H.E. Simon, *Models of bounded rationality* (The MIT Press, Cambridge, 1982)

X. Su, Bounded rationality in newsvendor models. Manuf. Serv. Oper. Manag. **10**(4), 566–589 (2008)

C.S. Tang, Perspectives in supply chain risk management. Int. J. Prod. Econ. **103**(2), 451–488 (2006)

A.A. Tsay, S. Nahmias, N. Agrawal, Modeling supply chain contracts: a review, in *Quantitative models for supply chain management*, ed. by S. Tayur, R. Ganeshan, M. Magazine (Kluwer Academic Publisher, Norwell, 1999), pp. 299–336

S.K. Vickery, J. Jayaram, C. Drodge, R. Calantone, The effects of an integrative supply chain strategy on customer service and financial performance: an analysis of direct versus indirect relationships. J. Oper. Manag. **21**(5), 523–539 (2003)

O.E. Williamson, *The Economic Institutions of Capitalism* (The Free Press, New York, 1985)

J. Yao, Decision optimization analysis on supply chain resource integration in fourth party logistics. J. Manuf. Syst. **29**(4), 121–129 (2010)

A. Zaheer, B. McEvily, V. Perrone, Does trust matter? Exploring the effects of interorganizational and interpersonal trust on performance. Organ. Sci. **9**(2), 141–159 (1998)

V.A. Zeithaml, A. Parasuraman, L.L. Berry, Problems and strategies in services marketing. J. Mark. **49**(2), 33–46 (1985)

About the Authors

Nicola Bellantuono is a Research Fellow in Operations Management at Politecnico di Bari (Italy). He holds a Laurea Degree in Management Engineering (2004) and a PhD in Environmental Engineering (2008). His main research interests deal with exchange mechanisms and coordination schemes for supply chain management, procurement of logistics services, open innovation processes, and corporate social responsibility.

Valeria Belvedere is an Assistant Professor in Production and Operations Management at the Department of Management and Technology, Bocconi University, and Professor at the Operations and Technology Management Unit of the SDA Bocconi School of Management. Her main fields of research and publication concern: manufacturing and logistics performance measurement and management; manufacturing strategy; service operations management; and behavioral operations.

Elliot Bendoly is an Associate Professor and Caldwell Research Fellow in Information Systems and Operations Management at Emory University's Goizueta Business School. He currently serves as a senior editor at the Production and Operations Management journal, associate editor for the Journal of Operations Management (Business Week and Financial Times listed journals). Aside from these outlets, he has also published in such widely respected outlets at Information Systems Research, MIS Quarterly, Journal of Applied Psychology, Journal of Supply Chain Management, and Decision Sciences and Decision Support Systems. His research focuses on operational issues in IT utilization and behavioral dynamics in operations management.

Stephanie Eckerd is an Assistant Professor at the University of Maryland's Robert H. Smith School of Business where she teaches courses in supply chain management. Her research uses survey and experiment methodologies to investigate how social and psychological variables affect buyer–supplier relationships. Her articles have appeared in the *Journal of Operations Management,* the *Journal of Supply Chain Management,* and the *International Journal of Operations and Production Management.*

I. Giannoccaro (ed.), *Behavioral Issues in Operations Management,*
DOI: 10.1007/978-1-4471-4878-4, © Springer-Verlag London 2013

Luisa Errichiello is a Researcher at the Institute for Service Industry Research (IRAT) of the National Research Council of Italy (CNR). She graduated in Management Engineering at the University of Naples "Federico II" and gained her PhD in Engineering and Economics of Innovation from the University of Salerno, Italy. She was visiting scholar at the Strategic Organization Design Unit of the University of Southern Denmark. Her research focuses on the microdynamics of innovation in organizations, the evolution of organizational routines and capabilities and the relationships between technology and organizational change.

Marco Formentini is a Research Fellow at Cass Business School in London. He graduated in Management Engineering at the University of Udine and completed his PhD in Operations and Supply Chain Management at the University of Padova. His main research topics are collaborative pricing and sustainability in supply chains, Product Development and Lean Design methodologies and Facility Management. He has published papers in *International Journal of Production Economics* and *International Journal of Production Research*.

Pauline Found is a Senior Research Fellow at Cardiff University. She joined the Cardiff University Innovative Manufacturing Research Centre (CUIMRC) when it formed in 2004 where she worked with the Lean Enterprise Research Centre (LERC) on sustainability of lean change. Since October 2007, she has been working in LERC on a number of research projects. She has published several papers on Lean and manufacturing improvement. She is a co-author of Staying Lean: Thriving not just surviving. She was President of the Production and Operations Management Society (POMS) College of Behavior in Operations Management 2009–2011 and a Fellow of the Institute of Operations Management (FIOM) and a Member of Chartered Institute of Purchasing and Supply (MCIPS).

Ilaria Giannoccaro is an Associate Professor in Supply Chain Management at the Politecnico di Bari, Italy. She graduated in Mechanical Engineering in 1998 and received her PhD in Management Engineering from University of Rome "Tor Vergata" in 2001. In 2011, she was visiting scholar at the Supply Chain Management Department of W.P. Carey Business School and joined the Research Center for Supply Networks (CaSN), Arizona State University.

Her principal research interests concern the management and organization of supply chain networks and geographical systems. She is author of more than 60 papers mostly published on international books and journals, among which *European Journal of Operational Research, International Journal of Production Economics, Journal of Geographical Systems, Production Planning and Control, Journal of Artificial Societies and Social Simulation, and Emergence: Complexity & Organization.*

Alberto Grando is a full-time Professor in Production and Operations Management at the Department of Management and Technology, Bocconi University, and Dean of SDA Bocconi School of Management. He is also a Visiting Professor of Operations Management at Cranfield School of Management, Cranfield University (UK). He has published a number of books and articles in academic and professional journals. His research interests are manufacturing performances measurement, supply chain management, and operations management.

John D. Hanson is an Associate Professor of Supply Chain Management in the Supply Chain Management Institute, University of San Diego. His research interests are in Knowledge Management and Innovation in the Supply Chain and the role of Behavioural Dynamics in Operations Management. He is an aerospace engineer by training and prior to his academic career he held executive positions with AlliedSignal (now Honeywell), Siemens and Eaton Corporation in the areas of advanced product development and technology planning.

Gregory E. Kersten is a full-time Professor of Decision and Negotiation systems and a Senior Concordia Research Chair at the John Molson School of Business, Concordia University. He received M.Sc. in Econometrics and a PhD in Operations Research from the Warsaw School of Economics, Poland. His research and teaching interests include individual and group decision making, negotiation analysis, decision and negotiation support, web-based system development, and electronic commerce. In 1996, he set up the InterNeg Research Centre involved in online training and development of e-negotiation systems, which since 2005 has been hosted at Concordia. He authored and co-authored six books, 25 chapters in books, and over 150 articles in journals and refereed conference proceedings; developed Web systems and interactive websites for teaching negotiations. He is a vice-chairperson of the INFORMS Group Decision and Negotiation Section, and senior editor of the *Group Decision and Negotiation Journal.*

Ugo Merlone is an Associate Professor at the Psychology Department, University of Torino, Italy He received his PhD in Applied Mathematics form University of Trieste, Italy. His main area of interest is the modeling of human behavior and organizations. On these topics, he has published articles on journals such as *European Journal of Operational Research, Physica A, Journal of Economic Behavior & Organization, Journal of Mathematical Sociology, International Game Theory Review, Organization Science, and Nonlinear Dynamics Psychology and Life Sciences.*

Rossella Moramarco received her PhD in Advanced Production System from Polytechnic of Bari in 2011. She has been visiting scholar at the R.H. Smith School of Business (University of Maryland) in mid-2010. She has been also visiting at the InterNeg Research Centre at the John Molson School of Business (Concordia University) in July 2010. Rossella's research interests focus on coordination of buyer–supplier relationships in the context of logistics services supply chains, sourcing mechanisms (traditional versus electronic negotiation), and behavioral operations management (e.g., trust in B2B relationships). She uses simulation as well as laboratory experiments methodologies. In April 2011 Rossella joined CEVA Logistics, where she works as transport engineer in 3PL and 4PL projects.

Pierpaolo Pontrandolfo is a full-time Professor in Business and Management Engineering at the Politecnico di Bari, Italy. In 2006–2009, he was the Head of the Department of Environmental Engineering and Sustainable Development. In 1995 and 1997, he was visiting scholar at the University of South Florida in Tampa, FL, USA. Since 2009 he has been collaborator of the InterNeg Research Centre, John Molson School of Business, Concordia University in Montreal, Quebec, Canada.

He is author of more than 100 chapters, mostly published on international books or journals, among which *European Journal of Operational Research, International Journal of Production Economics, International Journal of Production Research, International Journal of Project Management, Journal of Cleaner Production, and Journal of Product Innovation Management.*

Pietro Romano is an Associate Professor of Supply Chain Management, Product Development and Business Marketing at the University of Udine. He graduated in Management and Engineering and completed his PhD in Operations and supply chain management at the University of Padova (Italy) and was the founder and director of the first edition of the Master in Lean Management at the Cuoa Business School. His principal research interests concern supply chain management and lean synchronization. He took part in several research projects on lean- and SCM-related topics and is author and co-author of more than 110 publications. He has published articles in the *MIT Sloan Management Review, Supply Chain Management: An International Journal, International Journal of Production Economics, International Journal of Operations and Production Management, Industrial Management & Data Systems, International Journal of Production Research*, and *Interfaces*.

Boaz Ronen is a Professor in Business Administration at Tel Aviv University's Leon Recanati Graduate School of Business Administration. He has been visiting professor at several high-profile business schools, including Columbia University, New York University, and SDA Bocconi School of Management. He has published over 100 papers in leading academic and professional journals, and co-authored several books. His areas of interest involve Value Creation and increasing shareholders' value by using Value Drivers in areas such as Operations and Logistics, Information Technology, Sales and Marketing and Strategy.

Cynthia K. Stevens joined as the faculty at University of Maryland in 1990, after she received her PhD in psychology from University of Washington. Her research focuses on cognitive and social factors that affect individual and group behavior in the context of staffing (interviews, job search and choice, and recruitment), training (skill retention), electronic reverse auctions, and interpersonal work relationships (e.g., diversity dynamics, dysfunctional work relationships). Her work has appeared in *Personnel Psychology, Academy of Management Journal, Organizational Behavior & Human Decision Processes*, and *Journal of Applied Psychology*.

Josh Whitford is an Associate Professor at Columbia University Department of Sociology and is also a faculty affiliate at the Center on Organizational Innovation. In February 2007, he was named an Industry Studies Fellow by the Alfred P. Sloan Foundation. His interests include economic and organizational sociology, comparative political economy, economic geography, and pragmatist social theory. His research focuses on the social, political, and institutional implications of productive decentralization (outsourcing) in manufacturing industries in both the United States and Europe.

Francesco Zirpoli is an Associate Professor at the Department of Management of the Università Ca' Foscari, Venezia, and Research Associate at the MIT-

International Motor Vehicle Program. He gained his PhD in Management from the Judge Business School of the University of Cambridge, UK. and his Doctorate in Business Administration from the University of Naples "Federico II", Italy. His interests include strategy and organization design with a specific focus on organization boundary decisions, network governance, and the organization of innovation processes.

Printed in the United States
By Bookmasters